VCR Troubleshooting & Repair Guide

Robert C. Brenner
Gregory R. Capelo

HOWARD W. SAMS & COMPANY
A Division of Macmillan, Inc.
4300 West 62nd Street
Indianapolis, Indiana 46268 USA

WARRANTY WARNING

WARNING: Opening or otherwise modifying your VCR may void any manufacturer's warranty on the product.

WARNING: Dangerous voltages and currents are found in the VCR power supply. Only trained technicians should troubleshoot in or around power supplies.

©1987 by Robert C. Brenner, President, Brenner Microcomputing, Inc., and Gregory R. Capelo

FIRST EDITION
FOURTH PRINTING — 1988

All rights reserved. No part of this book shall be reproduced, stored in a retrieval system, or transmitted by any means, electronic, mechanical, photocopying, recording, or otherwise, without written permission from the publisher. No patent liability is assumed with respect to the use of the information contained herein. While every precaution has been taken in the preparation of this book, the publisher and author assume no responsibility for errors or omissions. Neither is any liability assumed for damages resulting from the use of the information contained herein.

International Standard Book Number: 0-672-22507-7
Library of Congress Catalog Card Number: 86-63297

Acquisitions Editor: *Greg Michael*
Manuscript Editor: *Malcak Associates*
Designer: *T.R. Emrick*
Illustrator: *Wm. D. Basham*
Cover Artist: *James R. Starnes*
Compositor: *Shepard Poorman Communications Corp., Indianapolis*

Printed in the United States of America

Trademark Acknowledgments

All terms mentioned in this book that are known to be trademarks or service marks are listed below. In addition, terms suspected of being trademarks or service marks have been appropriately capitalized. Howard W. Sams & Co. cannot attest to the accuracy of this information. Use of a term in this book should not be regarded as affecting the validity of any trademark or service mark.

Beta, Betamax, and SuperBeta are registered trademarks of the Sony Corporation.

VHS is a registered trademark of JVC.

Disney Channel is a registered trademark of Walt Disney Productions, Inc.

Sony is a registered trademark of the Sony Corporation.

Contents

PREFACE vii
    Growth VCR Market vii
    Troubleshooting & Repair viii
    Chapter Features ix

CHAPTER 1—INTRODUCTION TO VCR MAINTENANCE 1
    The First Cartridge Machine 2
    The Consumer Market 3
    Understanding the VCR 4
    Why Troubleshoot? 5

CHAPTER 2—BASIC TROUBLESHOOTING 7
    Introduction to Troubleshooting 7
    Steps to Successful Troubleshooting 7
    Component Recognition 8
    Component Failures 12
    How to Localize Failures 15
    When It Comes Time to Use a Service Center 19
    Safety Precautions During Troubleshooting and Repair 21
    Repair Parts 22
    Summary 22

CHAPTER 3—ROUTINE PREVENTIVE MAINTENANCE 23
    Extremes of Temperature 24

Moisture	25
Dust and Foreign Particles	25
Noise Interference	27
Power Line Problems	33
Corrosion	35
Magnetic Fields	37
Preventive Maintenance for Video Tapes	37
Preventive Maintenance for VCRs	39
Summary	47
CHAPTER 4—SPECIFIC TROUBLESHOOTING & REPAIR	**49**
Troubleshooting	49
When You Should Call the Repair Shop	50
Special Terms	50
Using Sams PHOTOFACTS	50
Audio Problems	52
Video Problems	64
Color Problems	71
Power Supply Problems	76
Improper or No Functions	79
Clock or Timer Problems	89
Summary	90
CHAPTER 5—MAGNETIC RECORDING THEORY	**91**
Video Head Construction	91
Magnetic Tape Recording Principles	93
Head Gap	98
FM Record/Playback	101
High Density Recording	106
Time Base Error	116
CHAPTER 6—VCR OPERATING THEORY	**135**
Automatic Color Control (ACC)	135
Automatic Color Killer (ACK)	137
The Bid Circuit	138
Current Chroma Signal Processing During Record	140
FM Modulator	147
Drop Out Compensation	151
E to E	156
Super Beta	156
VHS High Quality (HQ)	157
Servo Control	158
Early Beta Drum Servo	158
New Servos for Multiple Speed Machines	162

Separate Head Drum and Capstan Servo Responsibilities
  Defined 166
Understanding Microprocessors 177
Audio Recording Techniques 180
AFM Recording 184
Tuners 188
RF Modulators 193
Power Supplies 194
Summary 195

CHAPTER 7—ADVANCED TROUBLESHOOTING 197
Tools of the Trade 197
Components and How They Fail 206
Using Tools to Find Failed Components 208
Other Troubleshooting Techniques 210
Soldering Tips 215
Summary 218

GLOSSARY 219

BIBLIOGRAPHY 229

APPENDIX 231
Selecting a VCR 231
Choosing the Best Video Tape 234
Video Performance 235
Audio Performance 236
Preventive Maintenance Record 237
Routine Periodic Maintenance Schedule 238

*To our families, whose support and understanding
made this book possible.*

Preface

Seven-fifty for a movie? You've got to be kidding! Not really. Movie theater prices have increased steadily (almost in direct proportion to our kids' allowances). The five cent Saturday matinee of the late forties has become the four dollar "flick" of today.

However, an interesting phenomenon is occurring in the movie-watching consumer marketplace. Fewer people attend the expensive premiere showing, where long lines wrap lazily around the theater building. Instead, they prefer the comfort of home or apartment with a video cassette recorder (VCR) player, a large screen television, a rented movie, handy snacks, and drinks.

From the moment the first commercial home video player came on the market, the attendance at movie theaters began a slow, but steady decline. At the same time, the number of users who were purchasing their own VCR machines steadily increased. This increase became a torrent as VCR prices dropped from their original $1200 to less than $300 today. And standard features on machines today, such as wireless remote, multiday and multievent timers, hi-fi stereo, and direct cable capability were either not technically available ten years ago or simply too costly for the home video user.

GROWTH OF THE VCR MARKET

The video recorder market has steadily evolved over the last twenty years and is just now beginning to reach its prime. Today we can display fantastic color films on wide screens with ear-rumbling stereo sound. Stereo effects that could be experienced only in movie theaters are now possible at home. We can reproduce the same visual and sound effects from a rented copy of *Star Wars* that we saw and felt in the theater as a huge star cruiser roared out of the ceiling above and onto the screen. Sounds from a half dozen stereo speakers can fill our video world with the powerful engine rumble and action of this amazing space adventure. Today, far more films are seen at home on VCRs than in theaters, and this gap is widening.

In 1980, about 3 percent of U.S. households owned video cassette recorders. In 1985, over 28 percent, almost one-third of all U.S. households, owned a VCR! (The actual percentage may be less because many people own two.) In number of units, this equates to over 26 million machines in just the United States. In Europe and Japan, the number of home users is almost as high. The Electronics Industry Association (EIA) estimates that between 11.5 and 12.5 million VCRs will be sold in the United States in 1986 alone. A new VCR is being built every second. The number of home VCR owners is growing at an increasing rate.

A spokesman for RCA stated that "while two out of three households are still without the pleasures of home video, an ever-growing number of consumers are (now) buying their second VCR." Of the nearly 12 million VCRs sold in 1985, 2 million were second purchases ... second units or upgraded replacements.

Often an upgraded model can be purchased for less than the cost of an original machine. Users find a second VCR convenient for taping from one set while watching a program on another. Even more, if an outside antenna is used instead of a cable, the viewer can watch one program on TV, and tape a separate program on each of the two VCRs.

A second VCR is also indispensable for editing home videos and making copies of your masterpiece for relatives.

Finally, two VCRs let you rent tapes in either VHS or Beta format. However, nine out of 10 VCRs sold today are designed for VHS format tapes, so many movie rental stores are abandoning Beta and carrying only VHS format tapes. The recently introduced Sony enhanced SuperBeta VCR may reverse this trend.

TROUBLESHOOTING AND REPAIR

Owning a VCR is fun, it's exciting, and it's entertaining. But, when the machine begins to act strange or the picture on your TV starts looking weird, the thrill quickly fades and a form of recreational panic sets in. This stress reaches a peak when the VCR owner sets the machine on the repair center counter and learns that there is a sixty-five dollar charge just to open the unit and find the problem.

To make matters worse, the problem can often be traced to operator error or to some simple action that would have restored normal operation. Nowhere is the adage "Read the manual if you want to save time and money" more important and repeatedly confirmed than with the ownership and use of a VCR. Yet the manual is usually the first thing that a user misplaces or tosses away.

This book has been written to help all the users who either don't understand or can no longer find their operator's manual. It guides both novice and experienced VCR users through the magic of video recording. But more important, it gives valuable insights into those things that you can do yourself to restore correct operation or prevent failures from occurring.

CHAPTER FEATURES

Chapter 1, "Introduction to VCR Maintenance," provides a quick overview of the history of the video machine, and describes why VCR maintenance is so important. This preliminary chapter sets the stage for the rest of the book and shows both the novice and the experienced VCR user how each can gain maximum benefit from this book.

Chapter 2, "Basic VCR Troubleshooting," covers the basic steps in analyzing problems and repairing your own VCR. It describes the troubleshooting process and gives insight into VCR components.

Chapter 3, "Routine Preventive Maintenance," describes in depth the ways you can prevent the breakdown of your machine and preserve precious video tapes.

In Chapter 4, "Specific Troubleshooting," you will find detailed flow charts covering most VCR problems. These problems include: a screen that won't display; noise lines dancing across the screen; a VCR that won't record; and a VCR that won't play back.

Chapter 5, "Magnetic Recording Theory," assist those who wish to better understand the principles behind video magnetic recording. This chapter covers video tapes, the process of magnetizing the particles on the tape, and the techniques for playing back and reproducing the video and audio information stored on a tape.

Chapter 6, "VCR Operating Theory," extends the principles of magnetic theory to your video tape recorder. It describes the VCR components and tells how they interact to allow recording and reproducing both audio and video signals from tape. Both Chapters 5 and 6 were included in this book for those who wish more than a novice understanding of the VCR.

In Chapter 7, "Advanced Troubleshooting," you are introduced to the various types of tools and test equipment that become second nature to VCR repair technicians. Servicers of VCRs can use this chapter to familiarize new technicians with the equipment they will be using on the job.

The Appendix provides more meat-and-potatoes information related to the VCR. It covers selection criteria for VCRs and video tapes, a suggested periodic maintenance schedule, and a preventive maintenance checkoff chart. You will also find two comprehensive flow charts covering the playback and record modes of the VCR.

The Glossary explains over 200 terms and expressions common to video cassette recorder servicing.

VCR Troubleshooting & Repair Guide concludes with a comprehensive index that associates the subjects in this manual to a specific book page.

Acknowledgments from Robert Brenner

To Carol for keeping the house going while I was totally focused on writing this book.

To Jerome Mosley, professional photographer, for tough close-up and scope photos.

To Ed Roxburgh, San Diego artist, for his professional attitude and support in producing technical illustrations.

Acknowledgments from Gregory Capelo

To my wife, Debbie, who spent many hours typing this manuscript.

To Paul H. Huntington for the many hours freely given at the computer terminal.

To Dona Capelo and Vera Stern for assisting with my children so Debbie could type the manuscript.

To Lee Logue and David Neumann for technical support.

CHAPTER 1

Introduction to VCR Maintenance

Chapter 1 presents a basic overview of VCR maintenance important in learning how to troubleshoot and repair these machines. The intent of this chapter is to familiarize novices with the origin of the VCR and the internal parts of this magnificent machine, and to provide an introduction to problem and failure analysis.

"I still recall when I was a kid and my grandfather telling me that some day there would be an electronic device that could record pictures right off the television set. This machine would have the ability to play back both sound and picture immediately after it had been recorded. Then, during a later visit to his house, he proudly showed me his new Sony reel-to-reel half inch Portable Video Recorder Model AV3400. It was big and bulky, and was designed primarily for industrial use, but true to his word, there it was. It only recorded in black and white, and used large magnetic tape reels that held a half hour of program and had to be manually threaded in a fashion similar to an audio reel-to-reel tape recorder.

This marvel of that day included a black and white video camera and a special cable to record television programs. To record a TV program, he had to use a special monitor/receiver that could alter the TV signal so it could be recorded. It connected to the recorder through an eight-pin connector and cable.

The possibilities for this new device excited me. My grandfather then told me that someday in the not-too-distant future there would be a device that could reproduce color video and be able to record a TV program using its own tuner without even having the TV on." — Greg Capelo

For the past two years, futurists have been telling us that during our lifetime we will make three major purchases—a house, a car, and a computer. Today, this expression has evolved into "... a house, a car, and AN ELECTRONIC SUPPORT AND ENTERTAINMENT CENTER." The last purchase item includes the TV, the stereo, the computer, and the Video Cassette Recorder.

As consumers, we owe much to the broadcast industry for its persistence in developing video tape recording. About twenty years before the introduction of the AV3400, U.S. manufacturers such as Ampex and RCA busily worked on designs for video tape recorder (VTR) systems. Experiments were performed in 1951 to develop a rotating write–and–read head for recording and playback of video information.

In October 1952, the first barely visible video picture was demonstrated to an Ampex patent attorney and Ampex corporation founder Alexander Poniatoff. Also in that same year, Bing Crosby Enterprises demonstrated a broadcast VTR using fixed video heads and very high tape speeds. RCA introduced a longitudinal head VTR in 1953. It wasn't until January 2, 1955 that the first video cassette recorder (VCR) was produced using frequency modulation (FM) recording and special playback techniques. In that same year, a VTR using ½-inch tape was demonstrated in a closed circuit telecast from New York City to St. Paul, Minnesota. In 1958, sports history was made when the Los Angeles Rams football team used a video tape recorder to review the team's performance during a game half time period.

In 1960, Toshiba introduced a process called *helical scan* that resulted in smaller and lighter broadcast tape recorders. Two years later, RCA announced that it had developed the first fully transistorized VTR. But it wasn't until 1965 that Sony introduced the first consumer video (CV) VTR. This reel-to-reel machine used ½-inch wide video tape on reels that were 7½-inch in diameter. It could record one hour of black and white video using a technique called *skip field* recording. By skipping every other field of video information, one hour of recording time could be placed on a 7½-inch reel of magnetic tape. During playback, the recorded fields were played back twice reproducing an approximation of the original video image. But close observation revealed some loss of detail in the playback picture and a slight vertical jitter in the image.

Skip field technology did enable the recording of wider tracks of information on the tape which made it possible for the machine to tolerate some minor mechanical errors, with one major disadvantage. A normal TV cannot easily reproduce a playback picture without severe bending or flagging of vertical objects on the screen. Electronic synchronization of horizontal signals didn't work well when the same field was played twice so special monitor receivers were offered as optional accessories. These receivers had modified horizontal automatic frequency control circuits that produced a horizontally stable playback picture. These monitor receivers don't have their own tuner for channel selection.

Other manufacturers followed shortly with the introduction of new ½-inch reel-to-reel recorders. However, there was no interchangeability from one manufacturer to another. As new models were introduced by the same manufacturer, users discovered that tapes recorded on one model could not be played back on a different machine.

In 1968, Japanese manufacturers organized the Electronic Industries Association of Japan (EIAJ) and established a standard for ½-inch VTRs. These standards included mechanical and electrical specifications. Several manufacturers in Japan such as JVC, Matsushita, and Sony produced VTRs meeting these new specifications. EIAJ developed specifications for color recording about a year later.

The ½-inch reel-to-reel VTRs became very popular in industry, but the consumer market had not yet discovered home video recording.

Under pressure by the EIAJ, skip field recording was discontinued, and a process called *full field recording* was developed.

THE FIRST CARTRIDGE MACHINE

In 1969, Ampex designed a cartridge machine based on the Japanese color recording format. It used a plastic cassette to hold the magnetic tape reels, but it was never commercially manufactured. Then, in 1971, Sony introduced the U-Matic ¾-inch tape format mounted in its own cartridge or cassette. U-Matic capable VCRs were not sold to the public until 1972, and many Japanese companies bought licenses from Sony to manufacture tape recorders based on this new format. The ¾-inch U-Matic format is still widely used in industry, education, and government, but was too large and expensive for the home consumer market. The high cost of these recorders was the primary reason the consumer market remained extremely small.

In the early 1970s, an American company called Cartrivision, a division of Arco Industries, introduced a machine that used a cartridge containing two reels mounted one on top of the other. Using a modified form of skip field recording to conserve tape, the machine provided color picture information. This recording process skipped two fields for every single field it recorded (instead of skipping every other field like the Japanese machines did). Each recorded field was then played back three consecutive times. For example, field 1 was recorded, and fields 2 and 3 were skipped. Field 4 was recorded, the next two fields were skipped, and so on. During playback, field 1 would be played three times, then field 4 would be repeated three times.

Discontinuity in motion was visible in moving subjects, and this VCR required a modified TV to play back horizontally stable pictures. Cartrivision hoped to capture the consumer market by making available pre-recorded movies and sports events. These machines were actually test-marketed in Chicago by Sears & Roebuck, but poor sales caused the project to be abandoned.

THE CONSUMER MARKET

The consumer market for VCRs really began in 1974 when Sony introduced the Betamax home VCR in Japan. Beta-format VCRs were not available in the United States until almost two years later.

In the meantime, JVC introduced another format called VHS for "Video Home System." The VHS format received wide publicity and was adopted by many other manufacturers. VCR sales in Japan grew rapidly and soon Japanese VCRs were being sold in the United States. Zenith, Sanyo, Sony, and Sears began selling Beta-format machines, while Quasar, Panasonic, RCA, and JVC sold VHS machines.

Today, VCRs from many other manufacturers and from countries other than Japan are reaching American markets. Korean companies, for example, are manufacturing VHS machines designed to the same specifications as the EIAJ units and there are estimated to be over 30 million VCRs in the United States today.

The choice of format is mainly between Beta and VHS. Although an 8mm format was introduced in 1985, it has not been widely accepted.

Current Beta-format machines have three recording and playback speeds: Beta I (standard play), Beta II (long play), and Beta III (super long play). The early Beta I speed produced excellent video but limited recording time to about an hour and thirty minutes. Although Beta I was superseded by Betas II and III, it was recently reintroduced for the consumer who is especially conscious of video quality. Many of Sony's VCR products can play all three speeds.

Most VHS machines have three speeds: SP (standard play), LP (long play), and EP (extended play) or SLP (super long play). The length of recording and playback time is associated with the speed in which the tape passes through the machine. Faster speeds produce higher quality video reproductions but short duration recording capability. Slower speeds enable long duration recording (such as an entire baseball game) but experience a fall-off in picture and sound quality. Everything's a trade-off.

Table 1.1 compares the current video tape formats with recording and playback times.

Table 1.1. *Comparison of Video Tapes and Run Time*

BETA (L750 tape)	
Beta I (Standard Play)	90 minutes
Beta II (Long Play)	180 minutes
Beta III (Super Long Play)	270 minutes
VHS (T-120 tape)	
SP (Standard Play)	120 minutes
LP (Long Play)	240 minutes
EP (Extended Play)	360 minutes
SLP (Super Long Play)	360 minutes

The possibilities and benefits associated with VCR use are endless. Favorite television programs can be viewed at convenient times. Prerecorded classics can be rented or purchased and played at home. Special entertainment and educational programs can be recorded and viewed repeatedly. In fact, one particular exercise tape has sold more copies than *any* movie or classical tape. With the use of a video camera, family events can be recorded and a visual history produced for future enjoyment. Many owners of home movie films are now having their 8mm and 16mm films reproduced on video tape for enjoyment on a VCR.

UNDERSTANDING THE VCR

Connecting a VCR to a television in basic configurations is relatively easy. The hook-up is similar to connecting your TV to the antenna cable system in your home or business. Record and playback operations are much like those of an audio cassette recorder in a stereo.

By understanding the electronic and mechanical theory and operation of a VCR, you will develop a genuine respect for these marvelous machines and a sensitivity for proper care. You will also understand the machine's limitations.

The old expression "if you've seen one, you've seen them all" has relevance in the world of VCRs. In a sense the basic principles of operation have changed little since their invention. Only the techniques or methods change. The most basic VCR currently available uses the same principles for magnetic recording as the stereo hi-fi VCR priced three times higher. No matter what type or size box is used to package the recorder, the basic workings inside are common.

There must be an input section, a recorder section, a storage medium, a reader section, and an output section as shown in Fig. 1.1.

The *input* includes camera, microphone, another VCR, and cable and television signal paths. The *recorder section* converts all the input signals into a form suitable for saving, then writes or records that information on some form of storage medium.

The *storage medium* is the video tape itself. It's housed in a plastic cassette and comes in several sizes (formats). The cassettes are kept in special dust-preventing sleeves.

The *reader section* brings stored information, such as color television programs and sound signals, off the tape and converts it into signals suitable for the output unit. This output unit is actually the cable from your VCR to the television or to another VCR.

In all cases, the box, the television, and the "bells and whistles" (advanced features) can change, but the basic operation remains the same. A twenty-dollar audio recorder uses the same principles to record and play back as the video recorder near your television. The input signal from the microphone or broadcast station is passed from the input to the record section. Here the combined effort of electronic circuitry and mechanical parts positions the magnetic tape and then writes information magnetically on the top layer of the tape.

The same thing occurs in your VCR, but the electronic and mechanical components are far more complex and operating requirements (specifications) far more stringent. Your VCR must be capable of handling both video and audio signals. A great deal of information must be stored on the tape and tolerances are rigidly controlled.

The magnetic tape in the plastic video cassette housing is similar to the brown magnetic tape in an audio cassette. It's constructed from a long strip of plastic or mylar wound onto a reel. On one side of the tape, the manufacturer glues a material that can easily be magnetized, usually a form of iron oxide mixed in an adhesive binder.

In the recorder is a cylinder called a *drum* around which the tape is pulled. Tiny electromagnets called *heads* are mounted in the surface of the drum. Electric signals from the video and audio part of the recorder section are passed through coils of wire in these heads.

The heads produce a tiny magnetic field that changes in direct proportion with the changing au-

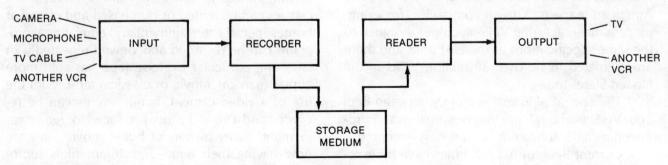

Fig. 1-1. The basic components in every VCR system.

dio or video signals. The heads make contact with the oxide surface of the tape passing around the drum causing the head magnetic field to be felt by the tape. The oxide on the tape becomes magnetized in direct proportion to the signal in the head. Since several heads are on the drum, several electric signals are converted to magnetic fields and applied to the tape, transferring the audio or video information onto the strip of tape where it's stored for later retrieval (playback).

When you insert a tape cassette into your VCR and begin to play back the stored information, complicated gears, arms, and other mechanisms pull the tape into the machine and place the tape against *read heads* on the drum cylinder.

These heads sense the tiny magnetic fields stored in the oxide layer of the tape and convert the changing magnetic fields (patterns of magnetic particles) into small electric signals which are amplified to make them strong enough to send out on the coaxial cable which runs from your VCR to a nearby television set.

The television set converts these electronic audio and video signals into light images on the screen and sound from a vibrating speaker system.

It's that simple ... in basic theory, at least. But to accomplish these feats, very complex circuitry and arrangement of mechanical devices are required. Tolerances are extremely tight, and signal levels are strictly controlled to enable excellent quality video from the surface of a shiny brown tape. Ensuring that these parts combine to reproduce brilliant video and ear-pleasing audio is the subject of the following chapters.

The VCR is a complex, precision machine. It's expected to operate correctly and without fail. Care and proper operation are mostly ignored while everything is functioning properly. But when the machine begins to fail from months of neglect and misuse, you really learn to respect it.

WHY TROUBLESHOOT?

One visit to a VCR repair center can leave a permanent impression on your mind and pocketbook. Most repair center costs average $125. Many service centers charge a flat fee which ranges between $120 and $140. The average labor charge is $45 per hour; typically two to three hours are spent troubleshooting and repairing a machine.

Some repair centers fix only the problem. Others also replace worn belts and do alignment checks to extend the operational life of the machine.

The value of a book teaching troubleshooting and repair for the novice is more evident when you understand the causes for typical VCR failures—dirt build-up and the wearout of belts and pulleys.

Users don't intentionally damage a VCR, but lack of knowledge and proper preventive maintenance can be costly. This book provides understanding and appreciation for the VCR so you can give it the tender loving care it needs. It will also help you identify problems quickly, fix failures, and get on with the show. VCRs can serve you well for many months; this guide will show you how and why.

CHAPTER 2

Basic Troubleshooting

Like automobiles, VCRs break down after lots of use. Some break down sooner than others. Finding the problem can be easy or difficult, depending on your understanding of how to analyze a problem, identify the failed part, and step toward the correct repair. This chapter will show you how to find problems in your VCR in the shortest amount of time.

INTRODUCTION TO TROUBLESHOOTING

Imagine for a moment that you're in the midst of watching a video tape when suddenly the TV screen goes blank and the VCR stops working. What do you do? What failed?

This chapter is devoted to something we often wish we could pass off or ignore—trouble. Trouble is like a flat tire: no one wants it, but when it happens we want to fix it quickly and get the experience behind us. Knowledge and action overcome trouble.

You know from reading the owner's manual that your VCR is an electromechanical machine; it operates using analog and digital principles to record and play back audio and video signals.

A VCR generally doesn't break down slowly, with graceful degradation (at least you can't see it). If it fails, it's usually with a hard, consistent failure. In addition, the electronic devices that make up your VCR function within strict rules of logic. The most effective way to respond to a failure in these devices is to think the problem through just as the machine operates, logically. Understand what should happen and compare the "shoulds," one by one, with what is really happening.

A deductive technique called *troubleshooting* is particularly appropriate for solving VCR failure problems. Troubleshooting could be really frustrating if you were left to struggle through the process by yourself. This book provides you with techniques for quick and easy troubleshooting and repair.

STEPS TO SUCCESSFUL TROUBLESHOOTING

Effective and efficient troubleshooting requires gathering clues and applying deductive reasoning to isolate the problem. Once you know the cause of the problem, you can analyze, test, and substitute (good components for suspected bad components) to find the particular part that has failed.

The use of special test equipment such as an NTSC pattern generator and an oscilloscope can speed the analysis, but, for many failures, good old brain power can suffice. Once you determine whether the problem is electronic or mechanical, deductive analysis changes to intelligent trial-and-error replacement. Reducing the number of suspected components to just a few and using intelligent substitution is the fastest way to identify the faulty device.

In general, follow these steps when your VCR fails:

1. Don't panic.
2. Observe the conditions.
3. Use your senses.
4. Retry.
5. Document.
6. Assume one problem.
7. Diagnose to a section (fault identification).
8. Localize to a stage (fault localization).
9. Isolate to a failed part (fault isolation).
10. Repair.
11. Test and verify.

The following pages discuss the steps to troubleshooting success in detail.

When something goes wrong, the first step is to determine whether the trouble results from a failure, a loose connection, or human error. Once you're sure a failure has occurred, the next step is to determine which section of the VCR is not operating—mechanical or electrical.

Then, step by step, partition each section into stages and try to track the trouble to a single component. If one function isn't working, for example, the problem could be in the activation button itself, the connecting wires, or the electronic circuitry of the VCR.

Next you need to understand what the machine is all about mechanically and how it interacts with the other parts of the unit.

COMPONENT RECOGNITION

What's a VCR made of? That housing or case is made of high-strength, molded plastic or metal depending on your unit. These cases are not likely to fail under normal use.

MAKE SURE THE POWER IS OFF, then open your VCR. Use the disassembly instructions found in the service literature.

There are a number of subassemblies inside your VCR including the VCR (circuit) boards, levers, gears, and tape drive motors, tuner, and power supply. Let's concentrate on the VCR boards, since this is where many failures occur.

To help you understand, refer to Fig. 2.1. Circuit boards, like the one shown in Fig. 2.1, are made of fiber glass and have many colorful items mounted on them. These items include sockets, connectors, wire traces embedded into the boards, integrated circuits (also known as ICs or chips), resistors, capacitors, and transistors as well as variable controls which align the unit.

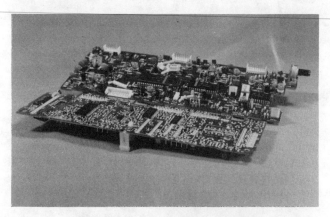

Fig. 2.1. A typical VCR circuit board.

Fig. 2.2 shows some of the devices you will find mounted on a circuit board.

Chips

Those black-case, centipede-looking things are the chips or ICs. They serve the function of hundreds of transistors (or vacuum tubes, the predecessors of transistors), and cause the VCR to work logically. Much of the unit works on the principles of digital logic. The system control works like a Von Neumann machine, a computerlike device that works in binary digits (bits). All conditions are ei-

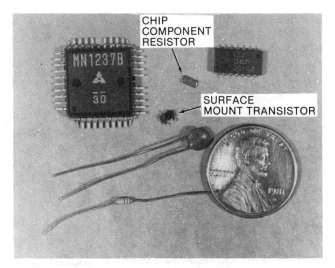

Fig. 2.2. Typical circuit board devices.

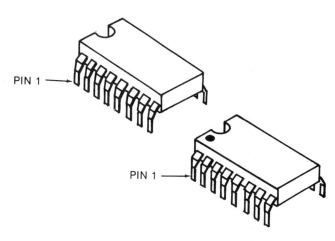

Fig. 2.3. Locating pin 1 on integrated circuits (chips).

ther ON (logic 1) or OFF (logic 0), and all operations occur in sequence. Dr. John Von Neumann first described his idea of a binary computer at a conference at the Moore School of Electrical Engineering in 1945.

The analog part of the machine operates more like a radio or TV. It receives signals, and amplifies and processes them for recording and playing back as needed. ICs are used in this part of the VCR too. They act to amplify, shape and switch the signals in the unit.

There are many sizes of chips on your circuit boards: 8-pin, 14-pin, 16-pin, 24-pin, 40-pin and 64-pin. Many of the ICs in a VCR are custom chips that are not inserted in sockets. So repair isn't quick and easy like it was in the days of tubes. In fact, special soldering skills are needed.

Notice that each chip has a notch or groove at one end as shown in Fig. 2.3. This notch marks the end of the chip where pin 1 can be found. Pin 1 is to the left of the groove as you look down at the top of the chip with the groove pointed away from you. The pins are numbered counterclockwise starting from pin 1, so that the highest-numbered pin is directly across from pin 1. As you'll learn later, in chip replacement, you must insert the new chip into the holes with pin 1 in the right place.

Chips have special markings that tell a lot about what's inside. Look at the printing on the top of the chips on your VCR's circuit boards. First, you'll notice that many different companies make chips, and that most of these companies are outside the United States—Korea, Japan and Taiwan, for example. Most companies place their logo on the chip.

You'll also notice letter-number combinations on the chips. Some chips have two sets of letter-numbers. One set identifies the type of device, and the other set tells when the chip was made.

The prefix ("MN" in MN1237B) usually identifies the manufacturer, although sometimes it identifies the device family associated with a manufacturer. The prefix is sometimes omitted.

The core number is three or more digits long. It indicates the basic logic family or the specific application in the circuit. Many of the chips in the VCR are specifically designed for applications *only* in that model.

Resistors

Fig. 2.4 shows three common types of resistors found in VCR circuitry.

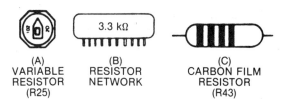

Fig. 2.4. Three types of resistors.

Resistors are used to restrict or limit the flow of electrical current through the board's circuitry. One type of resistor is the carbon film device shown in Fig. 2.4C. The value of resistance is given in ohms, and can be determined by comparing the color bands with the colors in Table 2.1.

Table 2.1. *Color Code Chart*

COLOR	DIGIT	MULTIPLIER
Black	0	1
Brown	1	10
Red	2	100
Orange	3	1000
Yellow	4	10000
Green	5	100000
Blue	6	1000000
Violet	7	10000000
Gray	8	100000000
White	9	1000000000
Gold		± 5% tolerance
Silver		± 10% tolerance

For example, imagine that the color stripes around the resistor are brown-gray-orange-gold, in that order. The first two bands describe the primary number (18). The third band represents the number of zeroes to add to the primary number—in this case three. The last band is the tolerance value, or how close to the color band value the actual value must be. As Table 2.1 shows, the brown band stands for 1, the gray for 8, and the orange for three zeroes after the 18. By using Table 2.1, this resistor's value can be found to be 18,000 ohms (18K ohms). The gold band represents a 5 percent tolerance value which means the actual resistance value can be plus or minus 5 percent away from the 18K-ohm designation.

Fig. 2.4B shows a single-in-line package (SIP) network resistor. Resistor networks are used by several components on the board at the same time. A recently developed electronic device, the resistor network is actually a group of resistors built into an SIP or a dual-in-line package (DIP). Several SIP resistor networks are mounted on the boards.

The resistance designation of network resistors is printed on the side of each package. Assume that one of these devices was marked 8X-1 2 0 2. The 202 is a key to the resistance value in ohms. The first two numbers (2 and 0) indicate the significant figures. The third number (2) tells how many zeroes to add to the significant figures. Thus, 202 means 20 plus 2 zeroes or 2000 ohms (2K ohms).

Some network resistors are marked directly with the value of resistance. One marking for example could be 898-1-R8.2K. The 8.2K labels this resistor package as a network of 8.2K-ohm resistors.

The device drawn in Fig. 2.4A is a variable resistor. These are used to align the electronics in the VCR. They are used throughout the unit to adjust the various circuits for proper operation. The tracking control on the outside of your unit is one example of a variable resistor.

Fig. 2.5 shows some carbon film and variable resistors. They are labelled on the board with an "R" prefix and a number.

Fig. 2.5. Closeup of a typical VCR circuit board.

Many different sizes of resistors can be found in your unit. The larger they are, the more heat (power wattage) they dissipate.

Capacitors

In addition to chips, circuit boards have a number of capacitors mounted on them. A capacitor builds up and holds an electric charge for a period of time determined by the design of the circuit. They are used to couple changing (A.C.) signals between circuit sections, to smooth ripples in waveforms, and to store potential for future use by other electronic components in the circuit.

Capacitors come in many varieties. Some of these include: (1) Electrolytic, (2) Tantalum, (3) Ceramic Film, (4) Mylar, and (5) Variable. These five types of capacitors can be found in the VCR circuit boards all over the unit. Fig. 2.5 also shows several different types of capacitors mounted on the board.

Capacitors are measured in fractions of farads. You'll see values listed in µF for microfarad and pF for picofarad. Micro means "to the sixth decimal place" or .000001 (one millionth) and pico means "to the twelfth decimal place" or 0.000000000001 (one trillionth). Thus .022 microfarad means 0.000000022 farad and 47 picofarad means 47 trillionths of a farad.

Capacitor value identification is challenging because most companies like to use their own identification standards. The designation "10µF" and "16v" on the side of an electrolytic capacitor means that this is a 10-microfarad capacitor rated at 16 volts.

A ceramic film capacitor with the value ".047 ± 10%" stamped on it is a .047 microfarad capacitor. Other ceramic film capacitors are color-banded and can be decoded using the color chart in Table 2.2. They have two additional color bands which refer to the capacitor's temperature dependence and tolerance. Capacitance varies with temperature (and to some extent with capacitor size and applied voltage). The temperature dependence of the capacitance value is given as the coefficient parts-per-million-per-degree-Centigrade (PPM/C). Tolerance defines how much variation is permissible between the actual value of the capacitance and the nominal capacitance value.

Table 2.2. *Capacitor Color Codes Indicating Capacitance in Picofarads*

COLOR	DIGIT	MULTIPLIER	TOLERANCE
Black	0	1	
Brown	1	10	
Red	2	100	
Orange	3	1000	
Yellow	4	10,000	
Green	5	100,000	
Blue	6	1,000,000	
Violet	7		
Gray	8		
White	9		10%
Gold			5%
Silver			10%

The electrolytic capacitor has a polarity marked on its case. Usually, the negative pole is indicated by an arrow. Electrolytic capacitors are shown in Fig. 2.5.

Variable capacitors are not common in a VCR. They are used most often in the adjustment of high frequency oscillations for the microprocessor timing clock pulse or in the chroma 3.58 MHz oscillators. A capacitor can be used to adjust the color signal. Variable capacitors are usually measured in the picofarads.

A mylar capacitor usually has a colorful, shiny, hard shell. Fig. 2.5 shows a mylar capacitor. It has 472j printed on it and underneath that, you can see 100e. This is a .0047 microfarad capacitor at 100 volts. The 47 represents the number of microfarads and the 2 represents the number of zeroes that precede the first two numbers. The "j" label represents the tolerance of the part capacitance rating.

Inductors

Fig. 2.6 shows two types of inductors found in a VCR. These are measured in microhenries. While we must determine the value of unmarked inductors by referring to the schematic, some inductor values can be determined by reading the colors on the device and comparing them with the color code chart in Table 2.1.

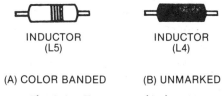

INDUCTOR (L5)   INDUCTOR (L4)

(A) COLOR BANDED   (B) UNMARKED

Fig. 2.6. Two types of inductors.

Diodes

Diodes are tiny, often glass devices shaped like resistors. They are marked with printing on the side, although you'll often have a tough time reading it. There are several diodes on the circuit boards of a VCR. To determine if you are looking at a diode or at something else, look for glass construction and a label on the side. A "1Sxxxx" label denotes a diode made in Japan. U.S. manufacturers identify diodes with a "1Nxxxx" label. Look at D509 located at the bottom right corner of the VCR board in Fig. 2.5. This is typical of many of the diodes found in your unit.

Transistors

The small black rectangular devices on the board in Fig. 2.5 are transistors. There's one in the lower left corner of the picture. It has three leads and is identified as Q512. Transistors can be found in different sizes and shapes in the VCR. The key to recognizing a transistor is its "2SAxxx, 2SBxxx, 2SCxxx and 2SDxxx" designators. Transistors made in the United States are labelled "2Nxxx." When you check the electronic parts catalogs, you'll find that transistors are not expensive.

Seventy-five percent of VCR problems result from mechanical failures, so unless you are experienced at replacement of electronic components, the introduction to chips, capacitors, resistors, diodes, and transistors will serve only to familiarize you with what is on your VCR circuit boards. These devices are soldered into the board and can be replaced only by those experienced in repair. Most home users will let a technician replace soldered-in components.

COMPONENT FAILURES

While the use of troubleshooting equipment is essential to analyze and isolate different VCR problems, many failures can be found without expensive equipment. In fact, troubleshooting and repair can be relatively simple if you understand how electronic components fail. Failures generally occur in the circuits that are used or stressed the most. These include the motor drive transistors and ICs, power supply transistors and mechanical switches. The microprocessors are highly reliable devices and seldom fail. Most electronic failures involve other components such as the transistors.

Often a VCR will exhibit strange symptoms when it starts to fail. The system control electronics will cause the unit to do things it never has done before, such as rewinding in the middle of a recording. The system control microprocessor is the first component suspected by inexperienced people who attempt repair. After spending hours trying to prove that the system control microprocessor is the failed part, a novice discovers that the end-of-tape sensor has failed and the system control microprocessor is doing just what it was designed to do at the end of tape movement—rewind.

Most manufacturers claim there are far more microprocessors replaced than necessary. The inexperienced technician often starts by replacing the microprocessor to correct a system control problem. After waiting for the part to arrive from the manufacturer, the technician replaces the IC only to discover that the problem still exists. Usually the culprit is a device which passes a status report to the microprocessor from another part of the machine. The only problem is that now someone has to pay for a microprocessor IC that costs $20 to $50. The manufacturer won't take it back because it's been installed. Guess who foots the bill for poor troubleshooting and repair.

Microprocessors do fail, but not often. The moral is to look for the cause elsewhere first and to check all signals, voltages and waveforms before jumping on the main system control IC.

Transistors and diodes can fail by internally opening (which causes an open or break in the circuitry), or by having a short in the output. Either situation causes total loss of signal.

Failures also result from *leaking*. A diode, for example, is designed to conduct current in one direction only. When it becomes leaky, it will allow some current to flow in the wrong direction. Transistors also leak under a partial failure operation. If either part leaks too much we say it has *shorted*.

Capacitors fail when they short internally or when one of the leads disconnects, causing an open. Again, there is a loss of signal.

Resistors can absorb too much current and actually bake in the circuit. The result is usually an open circuit with shorting during the "meltdown." All of the devices mentioned so far are solid state. They are constructed of materials (metals, plastics, oxide, etc.) that change as the components age or are subjected to severe temperatures or high voltages. Such a change can cause the device and the circuit or VCR to behave strangely.

Fortunately, VCR circuit boards are not normally subjected to high voltages. But they can get pretty warm if the natural air flow is restricted, and this affects the operation of the components. When we use the VCR, we place the circuitry under stress. It heats up when we turn it on and use it, cools down when we turn the machine off; and reheats when the machine is turned on again. This hot-cold-hot effect causes thermal stress. The result can be a break in the connection of a wire leading from inside the chip to a pin. This produces an *open circuit*, which requires chip replacement. Thermal stress can also create breaks in board traces and solder connections.

Even if there is no break in the chip or lead connection, after exposure to high voltages or temperatures the operating characteristics of a device can change. A chip may work intermittently or simply refuse to work at all. Theoretically, a wearout failure like this won't occur until after several hundred years of use. But we shorten the life span of the components by placing them in high-temperature, high-voltage, or power-cycling environments that cause them to fail sooner.

Other electronic problems occur outside the chip—between the chip leads and the support structure pins which connect the device to the rest of the VCR. Such failures include inputs or outputs shorted to ground, pins shorted to the +5-volt supply, pins shorted together, open pins, and connectors with intermittent defects. Most commonly, trouble results from opens or shorts to ground. Because they produce more heat, chips, transistors and diodes fail far more often than capacitors, resistors and inductors.

Chapter 3, "Routine Preventive Maintenance," tells more about the effects of heat. If you keep the VCR cool and clean, it should work well for many years.

Some electronic components fail more frequently than others. It might be helpful to know which parts to suspect first. You've already learned that devices that get warm, such as transistors, chips, and diodes experience thermal wear. These parts are high on the list of suspects. Also you should consider fuses, electronic sensors, and mechanical switches.

Defective capacitors would be next in the order of suspected components after the semiconductors (transistors, diodes and ICs). Last on the list of most likely suspects are inductors and resistors.

Electronic failures in a VCR most commonly occur in one of these parts.

- Mechanical switches and contacts.
- Fuses.
- Diodes, transistors, and ICs.
- Sensors- including hall effect ICs, light emitting diodes (LED), and photo sensors.

Second, you should consider these.

- Capacitors.

Finally, you should check these.

- Resistors.
- Inductors.

This grouping is by no means perfect, but it will provide some order of failure and tell you which parts to suspect first. The items within each grouping are not necessarily in order. Just because switches are listed first doesn't mean that they fail more often than ICs, but each of the items in the first list should be suspected before those in the following groupings.

Other Electronic Failures

Electronic failures during repair can be caused by careless mistakes at home, or in the shop by overzealous or undertrained technicians. Here are some possible repair-generated failures.

Devices "blown-up" in handling. This problem occurs when someone picks up static sensitive ICs like the microprocessor chip without first grounding any static electricity that a person might be carrying.

Bent or broken pins. Watch the way you put those chips in. You can only straighten a pin so many times before it breaks off completely (Fig. 2.7A).

Solder "splashes." These are caused by dropping tiny balls of solder from the end of the soldering pencil right on top of the board, shorting out some of the circuit as in Fig. 2.7B.

Liquid "fry." This occurs when someone holds or sets a liquid on top of or too close to the VCR and then accidentally spills the liquid into the top

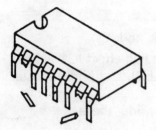 

(A) BENT AND BROKEN PINS   (B) SOLDER SPLASH

Fig. 2.7. Failures can be caused by overzealous or undertrained repair people.

of the machine while the VCR is running. It's a real mess to clean up, and you usually have to replace lots of components.

Component failure by asphyxiation. Blocking the VCR vent openings or stuffing your VCR into a tight cabinet where it has no room to breathe "kills" components.

The connector that doesn't. This can be an improper hookup or poor quality cables. Plugging cables into the wrong pin could blow some components. If cable connectors are badly corroded, or poorly soldered no signal can get through the cable.

The message is clear. If a mistake can be made, someone has probably made it.

How VCR Mechanisms Fail

Seventy-five percent of all VCR failures are a result of a mechanical defect or mechanical wear. Motors may not rotate at an even speed, mechanical levers may lose traction. Bearings wear and may become noisy or cause uneven rotation. Friction generated by pulling video tape coated with oxide particles along the tape path can cause guides and heads to wear. In time the video tape grinds its own track into the tape path causing tape alignment error.

Fig. 2.8 shows video tape wear on a Beta upper drum assembly. This wear was so severe that the tape would no longer travel correctly along its path.

Mechanical failures can appear to be a defect in the electronics section. A worn tape path, for example, can cause the audio to record and play back at very low volume levels. When the problem is diagnosed, the audio amplifier may seem defective. But after replacing several components in the

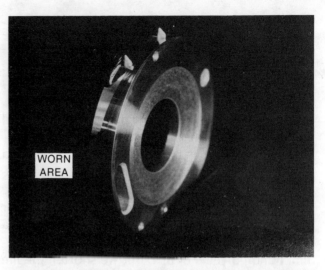

Fig. 2.8. Tape wear on a Beta upper drum assembly.

audio circuitry, you may discover that the signal coming from the audio head is insufficient because the tape is not traveling across the head as it should be.

Other mechanical defects may cause the system control to act strangely. Bent levers, which do not fully engage sensing switches can cause the microprocessor to misinterpret signals and operate strangely. The microprocessor depends upon information from many switches and sensors to keep the system control electronics informed of system status. If this information is missing from any of its inputs, the microprocessor may command the machine to shut down.

A defective capstan motor will cause the tape to be pulled through the machine at uneven speed. We hear this as *wow* and *flutter*—a wavering in the sound produced by the VCR. Music seems to warble in pitch and in severe cases even normal voice playback seems to flutter. The *servo circuit* attempts to correct this problem by making rapid voltage changes to compensate for motor imperfections, but the correction lags behind the errors generated by a defective motor. If you examine the servo circuitry with test instruments, you may suspect a defective electronic component when the problem is a mechanical defect with the capstan motor.

Isolating a defect within a VCR is easier if you can differentiate between mechanical and electronic failures. Mechanical devices fail much more frequently than electronic components, so most

VCR failures are due to mechanical defects or wear.

The mechanical aspect of a VCR is quite reliable if you operate the unit carefully and conduct periodic routine maintenance. But sometimes you forget. You operate the VCR in an area where someone taps cigarette ashes into a tray on top of the unit. Or you jam a cassette into the VCR and slam the cassette tray down. Rough treatment of a sensitive video machine causes rough display and operational performance.

One day you are in the middle of enjoying a rented movie and the machine quits. Not only does it stop playing the tape, but it refuses to let you have that expensive movie back.

HOW TO LOCALIZE FAILURES

Isolation of a VCR failure to a particular area involves localizing the defect to a mechanical or electronic malfunction. If the VCR will accept a tape, record, play back, fast forward, and rewind, your job will be easier.

There are two ways to localize failures; the technical (electronic) approach and the mechanical approach.

Technical (Electronic) Approach

This method uses several troubleshooting tools to measure voltage and waveform levels in the circuitry of the VCR. The tools include an audio test signal generator, an NTSC video waveform generator, an oscilloscope, a multimeter, a monitor/receiver, and the correct service literature.

The technical approach requires a knowledge of electronics, a basic understanding of how the unit functions, and test equipment. It is usually a last resort, so the technical approach has been reserved for Chapter 7, "Advanced Troubleshooting."

Mechanical Approach

An important part of mechanical troubleshooting involves careful observation of unit operation. Check for broken belts, bent levers, and foreign objects lodged in the mechanism. If the video cassette will not insert properly into the tray, carefully check that area for misaligned gears or bent levers. Fig. 2.9 shows a typical cassette tray. Check this for bent levers or misaligned gears.

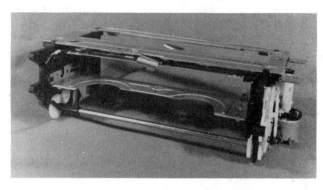

Fig. 2.9. Typical cassette tray assembly.

It's possible that an occasional plastic toy can mysteriously sneak into the tape load mechanism and wreak havoc on the mechanical operation. Fig. 2.10 is a typical mechanical chassis. Check for foreign objects that could obstruct tape or mechanical part movement, break belts or other components.

Fig. 2.10. A typical VCR mechanical chassis assembly.

If the VCR mechanism does not operate normally, a careful check of both the top and bottom portions of the mechanical chassis can provide helpful clues. Be sure to disconnect the power. Then expose the top and bottom chassis by removing the cabinet case and swinging out the circuit boards. If all appears to be normal, perform an operational check next.

Inspection of the mechanical components includes observing the operation of the machine during load, eject, play, fast forward and rewind operations. Be especially careful because the AC power is applied to the unit during these tests.

Use a video tape (with little or no value) to test for proper mechanical operation. Valuable tapes are never used for testing, especially if the mechanical condition of the unit is unknown. It would be a shame to have the VCR eat a priceless tape.

Most VCRs can be operated without a video tape in the unit by following the repair instructions in the manufacturer's service literature. End of tape sensors are defeated in VHS machines to prevent light from entering when no tape is in place.

The machine uses sensors called *cassette-down sensors* or *cassette-down switches* to tell when a tape is in place. The sensors can be defeated for troubleshooting purposes to make the unit think a tape is in place.

Once the VCR has been fooled into thinking that a tape is in the unit, you can engage the play, fast forward, and rewind functions. During this test procedure, check the unit for proper reel table, capstan, and head drum rotation as well as proper tape load and unload. If all appears normal, return the defeated tape end sensors and cassette-down switch to normal and place a tape in the machine.

Load the test tape into the VCR and attempt to play, record and fast forward. Carefully check the tape path to insure that it is not damaging the tape.

Check the condition of the video tape where each of the guides contacts the tape surface. Look for any creasing of the video tape at all points of contact. The tape is examined both where it enters and where it exits the capstan shaft. At this point a defective rubber pinch roller can damage the tape.

During this test, load and unload the video tape several times. Observe the video tape as the arms pull it out of the cassette during load and allow it to go back into the cassette during eject. These are critical operations that the machine must perform smoothly to prevent tape damage.

Inspect the levers and arms responsible for positioning the tape in the tape path during the various modes of operation. Be sure the placement is correct. These arms swing in and out of position at various times depending upon the function you select.

During play, forward hold back tension is measured at a point on the tape just before it enters the video head drum. Proper tape tension is critical. Too little tape tension reduces the picture quality; too much tension reduces the life of the video tape and increases the wear on the video heads.

If the VCR passes these checks, remove the tape again and fool the machine once more by defeating the end of tape and cassette down sensors. Measure fast-forward, rewind, and brake torques and check the play take-up torque. Improper torques indicate that cleaning or replacement of worn parts is necessary.

Some manufacturers sell test cassettes with built-in sensors that monitor and indicate the tension in the mechanical system. These cassettes are convenient but the price ranges from $150 to $300 per cassette.

VCR Troubleshooting Approach

Usually when an electronic or mechanical part comes to the end of its useful life, a catastrophic failure occurs. While you can't always see an electronic or mechanical defect, you can find the problem without much effort. (But, don't think that every time your VCR quits working, you've just had a catastrophic failure.) In most cases, the troubleshooting flow charts in this book will help you locate and correct the trouble quickly; but for problems that are not as easy to identify, follow these guidelines.

Don't Panic. You now have a manual that will help.

Observe. What conditions existed at the time of failure? What function was the unit in at the time of failure? What was on the TV screen? Did the VCR make any unusual mechanical sounds before it failed?

Sense. Is there any odor from overheated components? Does any part of the VCR feel overly hot?

Retry. If the VCR has no power, check the power plug and the power cord. Is the plug snug in the wall socket? Is the wall socket working? If not, correct the situation and try again.

If your unit has power, try ejecting the tape. If it comes out, replace it with a working tape (not a valuable one) and try to operate the VCR again.

If the unit has power, and appears to operate mechanically, reseat all the connector cables that may be associated with the failure symptom. Cables have a habit of working loose if they aren't clamped down. Once you've checked all the connections, retry.

If a tape won't load or unload, check the chassis for any obvious problems like broken belts or foreign objects. Repair techs often laugh about the times when they've lifted carrots, cereal, toast, doll clothes, toy soldiers, and coins out of the delicate mechanisms inside.

One woman chased an elusive cockroach into the VCR and promptly emptied a can of insect spray into the cassette tray opening. The cockroach died—so did her VCR. The spray destroyed the friction surfaces on the belt pulleys, and caused an acidic reaction that ate into the top layers of the electronic circuit boards. Getting rid of that single cockroach cost her over three hundred dollars (a new VCR and a replacement can of bug spray).

If the VCR has no power, unplug the unit and check the power supply for blown fuses. Fuses blow when too much current tries to pass through the wire. Once in a while they seem to blow on their own from age or some strange malady, but normally a blown fuse indicates a failure somewhere in the system.

Write. Document all that you see and sense. Write down all the conditions you observed at the time of failure amd the conditions that exist now that failure has occurred. Answer these questions.

- What is your unit doing?
- What is it not doing?
- What is being displayed on the function panel and on the TV screen?
- What is still operating?
- Is power still indicated by the clock and other parts of the VCR?
- Will the unit load and eject the tape?
- Is it damaging the tape?

Assume one problem. In electronic circuitry, the chance of multiple simultaneous failures is low. Usually, a single component chip malfunctions, causing one or more symptoms.

Diagnose to a section. You must determine if the malfunction is electronic or mechanical.

Consult the troubleshooting charts. Chapter 4 includes flow charts of the most common VCR troubles. If the symptoms that you see match a problem described in the charts, follow the instructions under "Troubleshooting Procedure."

CAUTION:
Any time you open the VCR, be sure the power is off, and touch a metal lamp, or other grounded object, to remove any stray static electricity.

Localize to a stage. Turn off the power to the VCR, and disconnect the power plug. Disassemble the VCR as shown in the service manual. Follow the troubleshooting steps and procedures in Chapter 4, "Specific VCR Troubleshooting and Repair," to localize the failed stage.

Isolate to the failed part. Closely following the procedures in Chapter 4 should guide you to the general area. But beware. There are many possibilities and potential pitfalls.

First, replacement of chips is not easy. Those fragile pins on the chips bend easily, and it takes very few straightening actions to break a pin completely off.

Removing and reinstalling electronic components that are soldered into the circuit boards are difficult actions and require more than a passing knowledge of soldering techniques. Only attempt this part of the test-repair procedure if you have experience soldering and desoldering delicate, printed circuit boards. Otherwise, bite the bullet and take your machine to someone who has the knowledge, experience, and equipment to successfully make this repair.

Sometimes a problem is caused by noise. Not audible noise, but electrical noise, the kind that produces static on the radio. This noise also affects VCRs. Noise in the VCR can cause very strange symptoms on the screen or in the sound. Fig. 2.11 shows how automobile ignition noise can cause interference in a TV presentation. Notice the noise dropout lines in the lower part of the screen and the jumping picture.

Fig. 2.11. Interference caused by automobile ignition noise.

> **NOTE:**
> To avoid noise problems, keep cables clear and away from power cords, especially coiled power cords.

And don't try out your new drill set next to the VCR while you're trying to set the unit to go off automatically. The VCR may decide to record the wrong program. The electrical current in your drill can affect the electronics of the VCR.

Sooner or later you're going to be confronted with those once-in-a-while failures called *intermittents*. These can be really frustrating.

Unlike a hard (constant) failure, an intermittent problem shows up randomly, or only at certain times (usually when you expect it least). Intermittent failures are difficult to locate using standard troubleshooting methods.

Since intermittent failures can be caused by shock, vibration, or temperature change, these conditions can be set up to find and correct the problem. Here are some helpful hints regarding intermittent failures. Be careful not to short out any connectors or pin leads. Be sure to use only a nonmetallic or wood object to probe components inside an energized VCR.

> **CAUTION:**
> The following steps are conducted with the VCR open and operating.

First, check, clean, and reseat all connector boards and cable plugs.

Second, tap gently at specific components on the suspected board using a nonmetallic rod or screwdriver.

Third, heat the suspected area with an infrared lamp or hair dryer. Don't overheat it.

Fourth, spray canned coolant on a suspected component. Service technicians sometimes use this technique to find a component that fails intermittently. Several companies sell pressurized cans of coolant spray that have long plastic extender nozzles for pinpoint application on top of a suspected chip. By cooling the device with the VCR energized, and operating the VCR, you can identify components that are on the verge of total failure. The VCR works for a few moments until the part heats back up and starts causing problems again.

Finally, after you've found the problem area, make sure the power is off, then use a strong light and a magnifying glass to look for small cracks in the wiring or solder connections.

If the problem is in a marginal part, replace the pesky rascal. Be sure your replacement is of the same type and specification as the original (i.e., replace CX193A with another CX193A).

A large section of Chapter 3 is devoted to identifying and solving the intermittent problem. For now, let's say that good cleaning, cable and board reseating, and inside-the-case temperature control will prevent the occurrence of most random failures.

The final method for fault isolation to a component is signal tracing. This technique will be covered in the advanced troubleshooting chapter (Chapter 4).

Removing and Replacing a Chip. A disassembly and reassembly guide is located in most service manuals. Look for yours before you begin.

It takes a little practice before you can remove a chip without it jumping out, flipping in midair and sticking you right in the thumb or index finger with that double row of tooth-like pins. An IC is the most difficult component to remove on any VCR circuit board.

But getting the chip out is only part of the challenge. Now you have to put the new chip in the board. Here's how to do it.

1. Replace all desoldered chips with chips of the same type and specification. Use a socket if it will fit on the board and the chip doesn't require soldering directly to the board.
2. Line up the pin-1 end (with the notch or dot) with pin 1 on the board.
3. Place the chip over the holes, lining up one row of pins with its holes.
4. With the chip at a slight angle, press down gently, causing the row of pins in contact with the board holes to bend slightly, which lets the other row of pins slip easily into their holes.
5. Press the top of the chip down firmly to seat the chip completely into the board. Be careful not to flex the board too much. If necessary, support the circuit boards with the fingers of your other hand as you press the chip into place.
6. Solder the chip into the circuit.

Now, that wasn't too bad was it? Even so, it is pretty easy to make mistakes in chip replacement. Here are two more tips to help you avoid Murphy's law.

Make sure you don't put the chip in backwards. The notch or dot marking the pin-1 end of the chip is intended to help you correctly line up pin 1 on the chip with pin 1 on the board and don't force the chip down in such a way that one of the pins actually hangs out over the hole or is bent up under the chip.

Test and verify. After the repair action, reassemble the VCR enough to power up and test the repair. This is important. You need to know that all is now well with the VCR.

> **NOTE:**
> It's a good idea to log the repair action in a record book to develop a maintenance history for the machine.

If these troubleshooting steps still don't help you find the failed component, you have two choices: either take the machine to a service repair center; or break out (or borrow) some test equipment, open the schematics, and start hunting for the failed or malfunctioning stage. Try signal tracing with an NTSC signal generator for the recording function. Use a tape with a known good recording on it for playback. Use an oscilloscope and a digital voltmeter (DVM) to test the discrete components such as transistors, capacitors, and resistors. Make voltage and resistance tests to locate the bad part. Test hardware and advanced troubleshooting methods are discussed in Chapter 7.

WHEN IT COMES TIME TO USE A SERVICE CENTER

You may clean the video heads and tape path and replace minor mechanical components, but there are adjustments and alignments required in the electronic and mechanical sections that should only be done by someone who has a good knowledge of VCR theory, the proper training, and the appropriate test equipment.

Although professional repair is costly, there are some things you can do to minimize the expense. So, before you call for help, read through the following list.

1. Identify the function that is affected.
2. Determine if the problem is mechanical or electronic.
3. Determine if the problem is caused by operator error.
4. Determine whether or not the problem is an intermittent failure.

5. Describe the problem in writing.
6. Carefully select a qualified servicer.
7. Record the serial number of your VCR.
8. Request an estimate.
9. Ask for a repair listing.
10. Determine if the repair is covered by warranty.
11. Get a receipt when you leave the unit.
12. Test the VCR before you accept the repair.

This checklist is a handy guide and should be used both before, during, and after the service center repair. Each step is expanded here for clarity.

Identify the function that is affected. Find out if the problem is catastrophic and affects all operations of the VCR.

Determine if the problem is mechanical or electronic. Be certain the problem isn't in the tape itself. Try to run several tapes you know are good.

Determine if the problem was caused by operator error. Be certain you didn't cause the problem.

Determine whether the problem is an intermittent failure. If the problem is intermittent, and you take it in for repair, it could be quite a while (at quite a fee) before the problem reappears and is fixed. You may just want to live with the problem until the intermittent becomes permanent (a hard failure). At least then you will have something concrete to troubleshoot. Intermittents are discussed further in Chapter 3, "Routine Preventive Maintenance."

Describe the problem in writing. Before taking your VCR to the shop, make a list of each of the symptoms it displays describing under what conditions the defect occurs. Is it intermittent or does it happen continually? Be sure each of the symptoms and complaints are written on the service order.

Carefully select a qualified servicer. Selecting a service facility that can give your VCR the proper attention is not as easy as picking up the phone book. It is much like seeing a doctor, and is important in prolonging the life of the VCR and video tapes. This is something that should be given careful consideration.

These suggestions will help you to select a servicer who will treat you fairly, professionally solve your problem, and help you avoid the frustration of dealing with incompetence.

- Do not assume that the most expensive is the best. The rates charged for labor do not necessarily correspond to the quality of workmanship.
- Check to see if the service facility has all of the required city and state licenses.
- Check the Better Business Bureau, Chamber of Commerce, and appropriate licensing bureaus (e.g., State Board of Repair) for any registered complaints.
- Ask if the service facility is authorized by the manufacturer to service your brand under warranty. Ask the manufacturer for recommendations of qualified servicers in your area.
- Determine if the servicer has been properly trained for VCR repair. Service centers often display certificates of training.
- Find out if they will be able to get parts on a timely basis. Most manufacturers have adequate parts available for servicers who deal with them regularly.
- Check to see if the service facility appears neat and orderly, the technicians and employees neat in appearance, the VCRs that have been checked in for service properly stored and not stacked on top of each other.
- Pay attention to the customer service representative. See if he seems knowledgeable about VCRs and concerned about your individual needs.
- Ask your local video rental club for recommendations. Ask friends, neighbors and video sales businesses who they recommend.

Record the serial number of your VCR. Jot down the serial numbers of all the accessories and the VCR you'll be turning over for repair. The service order should contain the manufacturer, model number, serial number, and a list of any accessories included. Also have the condition of the VCR cabinet noted on the work order when you check it in. Is it in good shape or is it all scratched up?

Request an estimate. Request a written estimate for labor and parts. Ask how long the shop will keep your VCR. Find out how the labor rates are determined. Some repair facilities charge on an hourly basis, others charge a flat rate.

If possible, ask one of the technicians what may have caused the problem, and what the solution may be. An experienced technician should be able to estimate required repairs but this does not mean the preliminary diagnosis will be l00 percent correct.

Ask for a repair listing. Ask for a detailed listing of repairs or replacements including a complete list of all charges.

Determine if the repair is covered by warranty. Make sure that parts and labor are warranteed for at least 30 days. What are the terms of the guarantee?

Get a receipt when you leave the unit. Get a copy of the work order showing the model and serial number of the unit, a list of accessories, the listed defects and the written estimate.

Test the VCR before you accept the unit. Test run the VCR before taking it out of the shop and again as soon as you get it home. Servicers who guarantee their work expect the unit to be operated during the guarantee period. But it is your responsibility to determine that the unit is in good working order before the guarantee expires.

SAFETY PRECAUTIONS DURING TROUBLESHOOTING AND REPAIR

As you would with any electrical device, you must observe certain precautions to prevent damage to yourself or the VCR. Observing these precautions can save you time, money, and frustration.

- Stay out of the power supply.
- Turn the power off, ground yourself against static electricity, and pull the plug.
- Handle video tapes carefully.
- Don't cycle the power quickly.
- Keep liquids away from the VCR.
- Handle components with care.

Don't troubleshoot the VCR power supply. These circuits convert the 115-volt line power in your home or office to the 5–12 volts used by the circuit boards. That 115 volts of electricity can be painful! Some VCR power supplies are connected to one side of the AC power line for their ground reference. If you touch this type of power supply without the proper isolation, you could get shocked. Limit power supply repairs to fuse replacement and only replace fuses with the AC cord disconnected. Be sure the replacement fuse has the same current rating as the old one.

Always turn the power off, touch a grounded metal object like a desk lamp, and then pull out the power cord before touching anything inside. Many failures are caused by people who don't follow this rule.

Handle your video tapes carefully. Don't lay tapes on a very dusty, dirty surface. Keep cigarette ash away from the tapes and VCR. Don't touch the tape surface. Don't set your tapes on, or in front of, a TV or color monitor.

Don't cycle the power on and off quickly. Wait 7 to 10 seconds so the capacitors in the power supply discharge fully and the circuits return to a stable (quiescent) condition.

Keep liquids away from the unit. It's amazing how sticky soda becomes after frying components all over the inside of the unit.

Handle components with care. Don't let chips lie around. The pins will get bent. Watch out for static electricity—chips may need special handling.

Special Handling

Some logic devices require extra care when you touch or handle them. The metal oxide semiconductor (MOS) chip family (MOS, CMOS, NMOS, etc.) needs some extra care since these chips are susceptible to damage by static electricity, so be sure to ground yourself by touching a metal lamp or grounded object before you reach for a chip inside the VCR chassis. In addition, conductive foam provides protection from static charge during storing or transporting of MOS-type chips.

Additional precautions to take when you use

test equipment with your VCR will be covered in Chapter 7, "Advanced Troubleshooting Techniques."

REPAIR PARTS

Finding that a trouble really exists is only part of the problem. You must locate the specific failure and then make the repair. This, too, can be challenging. Most of the chips in VCR boards are custom or special purpose parts and are not readily available at the local electronics store. You may be able to substitute the transistors for a generic brand, but the best way to replace any VCR part is with the original manufacturer's part type. The resistors and capacitors can be easily obtained at most electronic supply houses. Be sure to replace any critical safety components with only the manufacturer-recommended replacements. This is important for your safety.

The VCR chips and each of the other parts should be available from your local VCR repair service center.

SUMMARY

So there you have it. In this introductory chapter on troubleshooting, you've learned the troubleshooting steps to success. You've also learned how to recognize the components inside the VCR, how components and mechanisms fail, as well as various methods for localizing failures in your own machine.

CHAPTER 3

Routine Preventive Maintenance

In Chapter 2 you stepped through basic troubleshooting and corrective maintenance of the VCR. Chapter 3 discusses another type of maintenance, one that is intended not to fix a problem but rather to prevent a problem from ever happening. Preventive maintenance is in every way as important as corrective maintenance. In this chapter you'll learn what factors damage your VCR and cause it to fail, and what you can do to prevent these failures.

Often, the price you pay to buy your VCR is actually a small part of the overall system cost. The life-cycle cost of the equipment can be much larger than the initial purchase investment. This total cost increases dramatically as the costs for pre-recorded movies, books, magazine subscriptions, those extra connecting cables, blank tapes, and service center repair charges are added in. Service costs alone can grow to 10 to 50 percent of the system cost.

Occasionally we find a repair expense that exceeds the value of the broken equipment. It's when we look at high repair costs that terms like mean time between failure (MTBF) and mean time to repair (MTTR) become important. While your VCR has an excellent reliability track record, the way you operate your machine and the environment in which you place it become important to that MTBF number. Another factor to consider is that those bargain VCRs and blank tapes probably have less than an excellent reliability record. You get what you pay for.

Your VCR is sturdy, fun to use, and under most operating conditions, a very reliable machine. But, like other machines, it can wear out and fail.

As the conveniences of your VCR become increasingly essential in your home and business, your dependence on the machine increases. However, if repair is necessary, you may have to give up the machine for 1 to 3 weeks (although many problems can be fixed within a day).

Most large video production companies protect their huge investment in equipment because accidents and unnecessary failures can cost thousands of dollars in lost business. A small business, with a single video camera, two VCRs and other assorted equipment will experience just as catastrophic a loss by system failure, yet most don't take steps to prevent it.

VCRs don't burn out. They wear out or are forced out by human error or adverse operating conditions. If you misuse your VCR or don't protect it from environmental elements, you can be the cause for its failure.

A few moments of care can yield many more hours of good, consistent performance. We call this care preventive maintenance. Just as you periodically check the oil and water in your car's engine, and lubricate, wash and wax the body to keep it running right, so you should care for and protect your VCR. You can get good, reliable operation from your VCR for many months, if not years, if you provide timely and proper maintenance to keep the system in peak condition.

Proper preventive maintenance begins with an understanding of what we are fighting. Six factors can influence the performance of your equipment (not including the tape-eating dog or the cable-breaking baby). These factors are:

- Extremes of temperature.
- Moisture.
- Dust and foreign particles.
- Noise interference.
- Power line problems.
- Corrosion.
- Magnetic fields.

Each of these items acts to cause VCR breakdown. This chapter helps you successfully battle these enemies of reliable performance.

EXTREMES OF TEMPERATURE

The chips and other devices in your VCR are sensitive to high temperatures. During normal operation, the machine generates heat that is generally tolerable to the circuitry. Usually, leaving your VCR on for long periods won't hurt it because the slots and air vents let enough of the heat dissipate to the outside of the case.

As long as the components on the VCR boards and chassis are not too hot to touch, the amount of heat being produced should not cause any damage. However, heat can become a problem if you leave the dust cover on or confine the unit to a tightly enclosed cabinet during operation. Excessive heat within a component causes premature aging and failure. The heat produced during operation is not uniform across the device, but appears at specific locations on the chip (generally at the input/output connectors where the leads meet the chip itself). The usual effects of heating and cooling are to break down the contacts or junctions in the chip or other device, causing open circuit failure. When they are hot, these devices can produce intermittent operation of the VCR. The continual heating and cooling action during normal operation can also result in poor contact at solder points on the circuit boards.

Heat can contribute to cassette tape failure. Tape cassettes act just like audio cassettes when exposed to heat, especially the heat of the sun. If you leave cassettes in a hot car, some warpage will occur to the case or the tape itself will be damaged. If the tape is damaged too much, you will lose whatever is recorded on it. You could try to copy onto another tape, but the success rate for this "repair" isn't very high.

Heat is seldom a problem for the intermittent VCR user. When use is increased, generated heat can reduce performance and component lifetime. The following suggestions should help you prevent heat-related failure.

- Keep the dust cover off during operation.
- Allow plenty of ventilation around the unit.
- Keep the cooling vents clear.
- Keep the system dust-free inside and outside. Do regular preventive maintenance actions.
- Store cassette tapes in a cool, dry location.

The effect of cold on VCRs is interesting. The United States is currently working on superfast computers that operate supercold. Electronic components operate quite well in cold temperatures, but mechanical components have trouble functioning when the temperature drops. Consider the cassette tape drive mechanism for example. The operating range for a standard cassette tape drive is approximately 40° F to 115° F. At the low end, mechanical parts become sluggish and the possibility of improper tape movement increases. In addition, the tape itself can become brittle as it gets cold.

The rule of thumb for cold temperatures is to let the system warm up to room temperature (stabilize) before turning on the power. If the room temperature is comfortable for you, it's fine for the system.

MOISTURE

It's possible that humidity will build up and condense on the metal chassis parts and video head drum if the internal temperature of the VCR changes drastically. If the dew warning comes on after the unit is brought in from the cold, let it set at room temperature for a few hours. If you can't wait that long, use a hair blow dryer to dry out the machine. Condensation on the head drum causes the plastic tape to stick to the high speed rotating head drum. This can be disastrous to the tape and to the video heads. For this reason the dew sensor was designed to prevent operation of the VCR in the presence of moisture.

DUST AND FOREIGN PARTICLES

Just like flies at a picnic, dust seems to descend on VCR equipment. Static electric charge that builds up in the VCR and the television monitor attracts dust and dirt. That's why professional video tape editing systems are kept in cool, clean rooms. They require air conditioning and dust-free spaces, because the large equipment generates more heat, and is just as susceptible to failures caused by dust buildup.

Dust and dirt buildup coat circuit devices. This insulation blanket prevents the release of heat generated during normal operation. If the equipment can't dissipate this heat, the inside temperature rises higher than normal, causing the mechanical and electronic components to wear out faster. Dust is a major contributor to video head failure. Dust seems to be attracted to heat. Have you ever noticed that dust builds up on light bulbs in your lamps or on the tops of stereos and televisions more than it does on cooler objects? The charged dust particles become attracted to the magnetic field around electrical equipment. VCR problems increase in direct proportion to the increase in dust.

Mechanical devices like video cassette tape drives fail more often than solid-state electronic devices because mechanical and electromechanical devices have moving parts that get dirty easily, causing overheating and earlier failure.

Foreign particles such as dirt, smoke ash, and tiny fibers can cause catastrophic problems in the cassette cartridge and in the VCR itself. The air we breathe is full of airborne particles, but most of these are too small even to be seen, let alone become a problem. The larger particles in the air cause VCR problems. Cigarette ash, for example, can settle on the tape path and move from place to place inside the VCR creating improper tape travel and destroying record and playback heads.

Video tape sheds tiny particles as it moves through the machine. These particles prevent the heat generated during normal operation from escaping off the components and into the air. They also combine with the dust and dirt and become lodged in the lubrication material for the mechanical parts causing the moving parts to wear out faster.

Video heads and the tape path have more dust-related problems than the other mechanical moving parts. The tape must travel an exact path to record and play back properly. If dust is in the pathway, the tape can be lifted slightly as it passes. During operation, the video heads press slightly against the tape. As the head rides on the cassette tape surface, dust and dirt can cause major problems as suggested in Fig. 3.1.

Some current professional ½-inch video tapes are designed to minimize the attraction of dust and dirt by static electricity. This doesn't mean you can get careless about dust and dirt. Dirt on a tape can be swept off by the video head and gouge out a path on the cassette tape surface. It can stick to the head and cause other cassette tapes to be gouged, or cause the head itself to corrode and wear out.

Smoke from cigarettes and cigars can coat the internal surfaces of the cassette tape path with a gummy soot that not only produces tape path errors, but also increases wear by interfering with mechanical operation. Tobacco is also believed to cause rapid oxidation on pins and connectors, increasing the likelihood of intermittent errors. Also

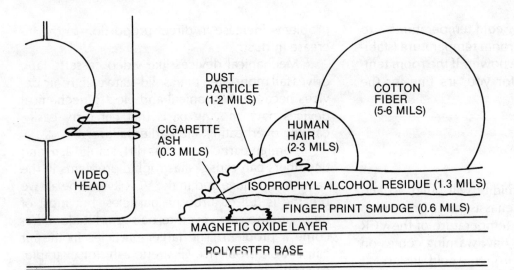

Fig. 3.1. Any small piece of foreign material can cause problems with a head that rides on the surface of the tape.

it increases the failure rate of rubber tires and belts in the tape drive mechanism.

Controlling Dust and Dirt

Dust buildup can be controlled. Thoroughly cleaning the VCR area every week will do much to keep your system in top condition. Dirt and dust can be removed from the equipment housings using a damp cloth lightly coated with mild soap. Clean electrical equipment with the power turned off. Be careful you don't wet or moisten the electronic components.

After washing the surface, rewipe the outside of the equipment with a soft cloth dampened with a mixture of one part liquid fabric softener to three parts water. The chemical makeup of some liquid softeners is almost the same as antistatic chemical spray. Wiping the cabinet helps to keep static charges from attracting dust to the top of the VCR. The fabric softener is antimagnetic and prevents the attraction of dust. The chemicals in this inexpensive solution can last longer than some antistatic sprays and help make your plastic VCR cabinet less susceptible to scratching.

Another successful technique involves blowing dust away from the VCR with a pressurized can of antistatic dusting spray. Using this product means you don't have to wipe off the equipment first. In any case, wiping a VCR should be done carefully, because you could scratch the paint, the plastic cabinet or the plastic windows if there are hard dust or dirt particles on the cabinet or cloth.

Here are some manufacturer-recommended cleaning methods.

- Use one part fabric softener to three parts water to clean indicator screens (including the television screen).
- Use mild soap and water with a soft cloth for drying.
- Use a window cleaner spray. (NOTE: Be careful in selecting the correct solution. Common household aerosol sprays, solvents, polishes, or cleaning agents may damage your cabinets and the TV screen. The safest cleaning solution is mild soap and water.)
- Use an antistatic spray.

CAUTION:
MAKE SURE THE POWER IS OFF AND THE PLUG(S) PULLED OUT OF THE POWER SOCKET(S). USE A DAMP CLOTH. DON'T LET ANY LIQUID RUN OR GET INTO YOUR EQUIPMENT.

Use compressed air to blow out dust and dirt from the VCR. Any brushes and vaccum nozzles poked into the unit may bump, scratch or damage the video heads or other critically aligned components.

You may not have an air-conditioned, air-purified room in which to use your VCR, so dust covers become very important. Plastic covers, made static-free with an antistatic aerosol or by wiping the surface with the fabric softener-water mixture, will provide good dust protection for your unit.

Don't place the VCR on the floor. Dust and dirt rise from the floor into the chassis increasing the potential for trouble. Use a cabinet or shelf. Here is a summary of ways to counter dust.

- Use dust covers.
- Keep windows closed.
- Don't operate the VCR on the floor.
- No smoking near your VCR.
- No crumb-producing foods near the unit.
- No liquids on any equipment.
- Don't touch the surface of any video tape.
- Vacuum around the unit area of your VCR weekly.
- Clean your VCR cabinet with static-reducing material.

NOISE INTERFERENCE

Your VCR is sensitive to noise interference, which can affect the proper operation or video picture. But what is noise, where does it come from, and how can you get rid of it?

Noise can be described as those unexpected or undesired random changes in voltage, current, picture, or sound. Noise is sometimes called "static." It can be a sudden pulse of energy, a continuous hum in the speaker, or a garbled display of characters.

Three types of noise cause problems: noise that affects the user (acoustic), noise that affects your VCR, and noise that affects other electronic equipment. For example, acoustic noise includes: the crying of a baby, the blare of an overpowered stereo, and the loud consistent tap-tapping of a noisy typewriter. Noise that affects the VCR and other equipment can be radiated, conducted, or received. It takes the form of electromagnetic radiation (EMR). EMR noise can be further classified as low- or high-frequency radiation (EMI or RFI), as shown in Fig. 3.2.

If the noise occurs in the 1 Hz to 10 kHz range, it is called electromagnetic interference (EMI). If it occurs at a frequency above 10 kHz, it is called radio frequency interference (RFI).

RFI can occur in two forms: conducted RFI, and radiated RFI. RFI that is fed back from the VCR through the power cord to the high voltage AC power line is classified as conducted RFI. In this case, the power line acts as an antenna, transmitting the noise interference out. When your VCR and its cabling transmit noise, this noise source is called radiated RFI.

EMI has three primary components: transient EMI, internal EMI, and electrostatic discharge (ESD). Transients occur when a large voltage pulse, or *spike*, smashes through the circuitry. Power line transients and electrostatic discharge from the human body are the two most severe forms of externally generated EMI.

Internal EMI is the noise generated within and by the chips, motors and other electronic devices. With current microelectronic designs, internal noise levels are so low that other factors such as connections and the length of leads have become the main source of noise in printed circuits. Internal noise does become a problem when the components are excessively heated or when the motors and components begin to fail.

The last form of EMI, the electrostatic discharge (ESD), is the same effect you get from walking across a carpet and then getting shocked by touching metal. ESD can cause the notorious "glitch" in electronic circuits.

All these types of noise interference can produce undesirable or damaging effects in the VCR. They can cause programs to stop in the middle of an operation, garbage to appear on the TV screen and chips to be destroyed. Noise interference must be prevented by reducing or eliminating its cause. This is not an insurmountable challenge, but it is a substantial practical and analytical task.

Origin of Noise Interference

Noise in the VCR can come from many places, including power supplies, fans, the VCR itself, the television, other equipment, connectors, cables, fluorescent lights, lightning, and electrostatic discharge. The lower portion of Fig. 3.3 shows how noise interference appears on a TV screen.

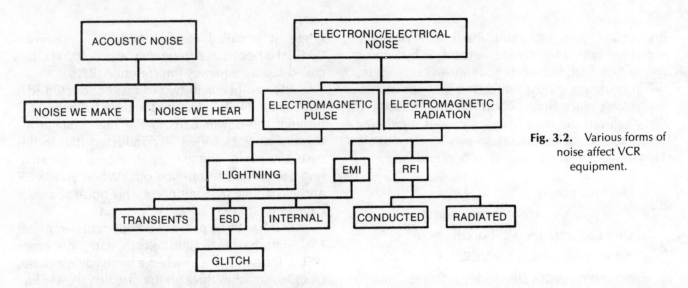

Fig. 3.2. Various forms of noise affect VCR equipment.

Fig. 3.3. Noise interference causes the white streaks visible in the lower portion of this TV image.

High-powered components used in switching power supplies has led to noise being conducted back into the power lines. Switching power supplies have been found to generate EMI in the 10–100 kHz frequency range.

Noise can even be passed or coupled to nearby equipment that's on a totally different circuit and not physically connected to our system. If two wires are placed next to each other, one wire can pick up signals coupled across from the other (crosstalk). Just 10 volts of electricity on one wire will cause a measurable (0.25 volts) on the other wire. Imagine how much crosstalk there could be if the voltage on one line were increased to 100 volts. The induced voltage on the other wire would be 2.5 volts, which is enough to change information in a stream of data being sent through that second wire.

Everything has some capacitance associated with it. Some typical capacitance values (in picofarads) are shown in Table 3.1.

Table 3.1. *Typical Capacitance Values*

SOURCE	CAPACITANCE (pF)
People	700.0
½-watt resistor	1.5
Connector (pin-to-pin)	2.0

Engineers have found that even 0.1 pF of capacitance can produce 5-volt spikes in digital circuits such as those found in the VCR.

Power line noise can feed into the VCR circuits whenever it exceeds the blocking limits of the power supply. Nearby high voltage machinery such as stamping mills, saws, air conditioning units, or clothes dryers can produce strong magnetic fields in the area around them and in their power cords.

Cables that vibrate and move in a magnetic field can also cause problems. Relays and motors can produce high voltage transients when they are turned on or off. And televisions and radios can be a source of noise coming into the VCR.

Finally, EMI can come from industrial, medical and scientific equipment, electric motors, home appliances, drills, saws, and tool speed controls. It's important to understand noise and how it can be generated. VCRs must be able to operate without causing interference with other nearby electronic equipment. They must be able to function without radiating noise, and in an environment that includes noise from outside sources.

Controlling Noise Interference

The most effective approach to noise reduction is prevention. If you can't prevent noise, you can at least take steps to minimize its impact. Here are five methods for dealing with noise.

- Filtering noise.
- Shielding from audible noise.
- Improved circuit board bonding connections.
- Improved wiring.
- Improved component design.

The most common approach is a combination of these methods, although filtering and shielding are most widely used.

Filtering involves the use of capacitors and inductors. There are many kinds of filters that respond to voltage, current, and frequency. For example, one filter prevents high-frequency voltage spikes from leaking out of a switching power supply into the circuitry being supported.

While most VCRs don't generate enough audible noise to require acoustic shielding or enclosure it may occasionally be necessary. If some form of audible noise reduction shielding is used, be certain there is adequate air circulation on the top, bottom, and sides of the unit.

Electromagnetic Interference (EMI)

EMI is an unplanned, extraneous electrical signal that affects the performance of the VCR system. It can cause automatic timer errors and unusual functioning of the unit. It can appear as circuit crosstalk, voltage ripple, or power supply drift.

While circuit designers try to minimize EMI, it is a natural by-product of aging components, bad solder joints, damaged or corroded connector contacts, and loose connections. It is also produced when a burst of electromagnetic or electrostatic energy is conducted or induced through the circuitry. Externally produced EMI enters the VCR through the cabling or openings in the case.

Some VCR cases are made of metal that is lightweight, durable and generally rustproof. While these are good qualities, the most important feature of these cases is that metal conducts electricity away from sensitive components so it provides protection against EMI/RFI and even ESD noise. VCR design engineers have tried many techniques to reduce EMI/RFI.

- Use decoupling capacitors (0.01 μF to 0.1 μF).
- Carefully lay out components.
- Keep traces as short as possible.
- Shield sensitive circuits.
- Reduce noise sources.
- Use fewer components.
- Carefully route wires.
- Use a shielded cabinet with minimal openings.

Cables are a source of EMI/RFI. Both the internal cables and those external cables leading to the TV, and antenna or cable system can radiate interference.

You can improve on the VCR's design to counteract EMI. Since you won't be changing the circuit board design, you reduce EMI interference in two ways: (a) by preventing it from reaching the circuits, and (b) by keeping it contained within shielded enclosures. To do this, use shielding, grounded cables, filters, and transient absorbers.

A shield is a conductive coat or envelope placed around a conductor wire or group of wires to provide a barrier to electromagnetic interference. EMI-reduction devices include the ferrite "shield beads" which are placed on power supply leads, connections to ground, or between stages on the circuit board.

Ideally, the shielded connectors should provide a continuous shield from the device, through the connector, and into the shielded cable. Otherwise, the weak shield point becomes a transmission hole for EMI to get out and interfere with other devices or appliances in the area.

One excellent countermeasure to EMI and RFI is the use of fiber optic cables and connectors. This technology hasn't yet become popular with video because the cost is still too high, but the day is coming when fiber optic video transmission will be the norm rather than the exception.

Electrostatic Discharge (ESD)

It sometimes appears that a wizard with a weird sense of humor secretly loads into every VCR timer a program that intermittently produces random glitches to drive the user wild. Chasing and catching the elusive phantom glitch is a challenge even for experienced repair technicians using expensive and complex troubleshooting equipment. But you can learn how to prevent this intermittent problem from affecting your VCR operation.

Glitches are short-lived electrical disturbances that often exist just long enough to cause problems in digital circuitry. They are usually the result of an electrostatic discharge (ESD), also one of the most severe sources of EMI.

People and objects can accumulate a substantial electrical charge or potential. The human body can accumulate static charges up to 25,000 volts, and it's not unusual to build up and carry charges of 500 to 15,000 volts. Charged objects or people can then discharge (quickly get rid of the voltage) to a grounded surface through another object or person. Remember the times you dragged your feet across the carpet and then shocked someone nearby? This electrical charge is static. It can discharge through your VCR, and when it does, all sorts of undesirable things can occur. If a program is running and a VCR user carrying a large potential of electrical charge touches a function key on the unit, the arc of discharge will find the shortest route to ground, usually through the sensitive microelectronic ICs, and the VCR will stop. If the VCR has a screen display showing the particular mode of operation or the automatic timer program shows on the TV, it can go wild and display weird characters. Sensitive components can be damaged or destroyed. Even a charge of only three volts is enough to create erroneous information in most logic circuits.

Electrostatic charges can be of any voltage. Here are some of the sources of ESD glitches.

- People in motion.
- Overheated components.
- Improper grounding.
- Poorly shielded cables.
- Improperly installed shields.
- Missing covers and gaskets.
- Circuit lines too close.
- Poor solder connections.
- Low humidity.

We know that static occurs when two objects are rubbed together. Your movement, walking while wearing wool or polyester slacks, can cause a tremendous charge of electricity to build up on your body. When this charge reaches 10,000 volts, it is likely to discharge on any grounded metal part.

Litton Systems, Inc. has developed the *Triboelectric Series* chart shown in Table 3.2. Cotton is the reference material since it absorbs moisture readily and can easily become conductive. If

Table 3.2. *Litton Systems, Inc. Triboelectric Series*

	Air
	Your hand
	Asbestos
	Rabbit fur
	Glass
	Your hair
	Nylon
	Wool
	Fur
	Lead
	Silk
	Aluminum
	Paper
Reference material*	Cotton
	Steel
	Wood
	Hard rubber
	Nickel, copper
	Brass, silver
	Gold, platinum
	Acetate, rayon
	Polyester
	Polyurethane
	Polyvinyl chloride
	Silicon
	Teflon

*When a material above cotton is rubbed with a material below cotton, the material above will become positively charged and the material below, negatively charged.

any material on the list above cotton is rubbed with any material on the list below cotton, the item listed above will give up electrons and become positively charged. The item listed below cotton will absorb electrons causing it to become negatively charged.

The two oppositely charged materials will tend to cling together. If they are separated, a static charge difference occurs. If teflon is rubbed in your hands, a large electrostatic charge is built up. The farther apart the materials are listed in Table 3.2, the larger the charge that can build up. Notice that hair is listed above cotton. Hard rubber is below cotton. Paper is listed just above cotton. Have you ever pulled a rubber or plastic comb through your hair and then used the comb to pick up pieces of paper? This is electrostatic charge in action.

A problem occurs when this charge builds and becomes quite large. Just walking across a carpet can generate over a thousand volts of charge. If the humidity is low and the air is dry in the room, the charge can be substantially higher. When the relative humidity is 50 percent or higher, static charges generally don't accumulate. However, in a dry room, built-up static charge will readily arc to any grounded metal, such as a VCR chassis.

An ESD release on the chassis case won't hurt you, but it can be very damaging to the electronics. The discharge pulse drives through the case to the microelectronic circuitry where it can burn out some of the chips. Even if no components are "fried," the damage that is done by this overvoltage spike accumulates and starts to degrade the functioning of some circuit board components. ESD damage costs have been estimated at millions of dollars annually. This figure is even higher when we consider components that are not totally destroyed, but are only degraded. Sooner or later the degraded components fail completely.

In low humidity, walking across a synthetic carpet can charge your body to 35,000 volts. Walking over a vinyl floor can charge you with 12,000 volts. A poly bag picked up off a table can develop 20,000 volts. Even sliding off a urethane-foam padded chair can load you with 18,000 volts. Is this a hazard to your VCR? Yes. As Table 3.3 describes, some electronic devices are very susceptible to low values of ESD voltage.

Table 3.3. *Voltages That Can Damage Electronic Devices*

DEVICE	DAMAGING VOLTAGE (MINIMUM)
CMOS chips	250–3000 volts
Diodes	300–2500 volts
EPROM memory	100 volts
Operational amplifier	190–2500 volts
Resistors	300–3000 volts
Schottky (S, LS) chips	1000–2500 volts
Transistors	380–7000 volts
VMOS chips	30–1800 volts

If your VCR occasionally gets the "shock treatment" or functions erratically, there are some things you can do. The following list offers some specific solutions to ESD problems.

1. Apply antistatic spray with a soft cloth on rugs, carpets, and VCR equipment to reduce and control static.
2. Install a static-free carpet in your VCR area.
3. Mop hard floors with an antistatic solution. The antistatic floor finish works well for up to six months, but this is an expensive solution more suited to electronic manufacturing facilities.
4. Place your VCR on a conductive table top.
5. Use static-free table mats.
6. Keep chips in conductive foam (that black styrofoam-looking material).
7. Touch a grounded metal object before touching the VCR.

You can defeat ESD glitches by paying attention to static charge in and about the VCR. By making static charge elimination part of a preventive maintenance program, you help extend the life of your system.

Radio Frequency Interference (RFI)

Radio frequency interference noise is much the same as EMI except it occurs at higher frequencies (greater than 10 kHz). RFI is what causes five other garage doors on your street to open when you operate your new automatic garage door opener. You'd really rather it didn't work like that.

Although RFI isn't a health hazard, it can be a

real bother. Many RFI problems are caused by placing the VCR too close to the television set. Sometimes the high frequency signals generated inside the TV interfere with the VCR's carefully timed signals. This usually occurs when the VCR is placed directly on or under the set. The trouble disappears when the unit is moved to the side or up on a shelf a few feet above the TV.

Another common source of RFI is the location of the nearest radio station transmission antenna. The color-under frequencies in your VCR occur around 600 kHz. It is not uncommon to have a VCR cause color rainbow-like patterns on the TV screen when playing back a color program if you live near a transmitter that operates on a frequency close to 600 kHz (60–68 on the AM radio dial) due to some harmonic (doubling or halving) of that frequency. The color signals inside the VCR can mix with the RFI and cause herringbone patterns like those in the test pattern in Fig. 3.4.

Fig. 3.4. RFI can cause the tweed or herringbone pattern to be produced on a TV screen.

The rainbow effect can appear as vertical, horizontal, or diagonal color stripes, or as red-green-blue herringbone or zig-zag shapes. This particular problem is normally seen in the playback mode. You can verify the cause by following these suggestions.

First, check to see if you live near a transmitter at or near 629–688 kHz (or some harmonic of these frequencies).

Second, when the problem appears, turn the color control on the set all the way off. If the pattern goes away when the color is turned off, you can assume that RFI from an interfering radio station is the cause.

Third, place the unit in PLAY and completely wrap the VCR in tin foil. If the trouble disappears, the RFI is from an external source and not internally created. (NOTE: DO NOT LEAVE THE UNIT WRAPPED IN THE FOIL OR IT WILL OVERHEAT.)

Finally, operate the unit in playback when you know the radio transmitter is off the air. (This may be very early in the morning.) If the tape plays without the rainbow interference you have isolated the problem.

A third form of RFI is unique to the audio circuits in a Beta hi-fi or a VHS HD audio unit. The cause is similar to the color rainbow interference just described, but the frequencies and symptoms are different. Beta hi-fi and VHS HD are recorded and played back on the video tape using FM frequencies ranging from about 1.3 MHz to 1.83 MHz. Sometimes when a unit operates near transmitters which broadcast in this frequency range (e.g., in the upper AM radio band), the VCR may receive and record or play back the station along with the taped program. You'll notice this symptom when you play back the recorded information and hear the interfering radio station signal. You can isolate this problem by following these steps.

First, confirm that all connecting audio cables have a good ground connection. Then, check to see if you live near a transmitter broadcasting in the range of 1.3 to 1.83 MHz (or some harmonic of those frequencies).

Next, place the VCR in the mode in which the symptoms occur and completely wrap the unit in tin foil. If the interference goes away, the RFI is external (not internally created). (NOTE: DO NOT LEAVE THE UNIT WRAPPED IN TIN FOIL OR THE UNIT WILL OVERHEAT.)

Finally, operate the VCR when the transmitter is off the air. If the symptoms disappear you can assume that the transmitter is the source of the problem.

Other RFI problems find their way into the FM video circuit preamplifiers. The symptoms can vary, but usually they are caused by nearby transmitters that interfere with the video signal in the first stages of amplification.

The only sure way to completely block RFI from getting into the VCR is to completely enclose it in a shield. Some manufacturers have designed special lead covers which provide suitable ventilation for the unit while it's operating and stop most external RFI troubles. In some cases these covers are expensive but more often they are simply not available due to limited quantities. Many manufacturers are now using all metal cabinets instead of the popular plastic cabinets common to earlier VCRs.

There are other ways to minimize or reduce RFI problems. For example, a smaller number of components in a system reduces the number of sources for RFI and improves system operation. Reliability improves in direct proportion to RFI improvements. Here are some more ways you can improve your system's RFI condition.

- Locate your VCR at least a couple of feet away from any television set.
- Use a directional outdoor TV antenna.
- Subscribe to cable TV.
- Connect traps or line filters to your VCR.
- Replace the antenna twin-lead wire with 75-ohm coaxial cable.
- Use only good quality audio and video cables to connect external equipment.

POWER LINE PROBLEMS

One important environmental factor for your VCR is good, clean power. If you depend on the local utility to supply this power in steady, reliable consistency, you may be disappointed.

While room lighting systems can tolerate line voltage problems that momentarily dim the lights when a large power-hungry machine turns on, VCRs cannot. Your unit, like most electronic units today, is more sensitive to power line disturbances than other electrical equipment. Even well-designed machines such as yours are affected by the quality of power provided. Under voltage or over voltage puts severe stress on VCR components. The effect is to accelerate the conditions under which a device gradually weakens, becomes marginal, and finally wears out. There are four types of power-line problems that cause concern:

- Brownouts.
- Blackouts.
- Transients.
- Noise (discussed previously).

Brownouts

Brownouts are those planned (and sometimes unplanned) voltage sags, when less voltage is available to drive your power supply, motors and solenoids. Brownouts are far more common than you may realize.

Voltage dips are common if you operate the unit near some large electrical equipment such as air conditioners or arc welders. Line voltage can be drawn down as much as 20 percent by the heavy momentary drain caused when this equipment is turned on.

The VCR should still work with line voltages that drop and remain 10 to 20 percent below the 110 volt rating. But if the supply voltage gets too low, the regulators in your power supply won't be able to pump adequate power into the unit. During brownouts, VCRs can operate intermittently, overheat, or simply shut down and lock up.

By the way, your power supplies can also handle a voltage "brown-up," or increased line voltage, but the power supply regulators will generate a lot more heat as they handle the extra incoming voltage.

Blackout

Power line blackout, a total loss of line voltage, can be caused by storms and lightning. It can also be caused by vehicles accidentally knocking down power lines or by improper switching action by a power station operator.

When power is lost, whatever you were recording may be gone. If you are operating in the automatic record mode when power fails, you will have recorded only a partial program. The part of the program being broadcast while the power was out (as well as anything after that) is not recorded. Many VCRs have a battery back-up built into the system control that keeps the clock circuitry running and time activations will save anything you have programmed into the automatic timer. These batteries keep the correct time on the clock and last for up to an hour. The unit will not have a display until the AC is restored from the utility company but the clock and timer are still alive.

The VCR will not turn on at the time programmed in the auto timer if no power is available, but will wait for the next timer-scheduled program to turn on the machine. Not all VCRs have this battery back-up feature.

If the weather turns bad and thunder is echoing across the sky, don't turn on your unit. If a blackout occurs or if you see lightning, turn off your machine, pull the plug, and disconnect all antenna leads until the storm passes.

And when the power goes out, be careful. While the room lights are out and you're muttering under your breath as you feel around for a flashlight, remember what is sure to happen when power is restored—a tremendous voltage spike will be produced as lights and motors go on all over the neighborhood. This could damage the VCR system. Always unplug your unit when a blackout occurs. Wait until power has been restored for a few minutes, then turn the system back on. Don't subject the power supply filters to these kinds of spikes.

The antenna leads should be disconnected to prevent any damage to the tuner, RF modulator and electronic TV/VCR switching devices that could be caused by lightning or falling power lines contacting the incoming antenna wire or cable. Even lightning that strikes a community cable system wire several miles away can damage your unit. Many owners now wish they had disconnected the antenna leads before the storm arrived.

Transients

Other than electrostatic discharge, power line and antenna line transients are the most devastating form of noise interference in electronic circuits. Transients are large, potentially damaging spikes of voltage or current that are generated in the power lines that feed electrical power to your community. Spikes can be caused by lightning, by failure of utility company equipment, or by the on/off switching action that occurs when you use any electrical tool or appliance.

Most of these spikes are small and barely noticeable. But some voltage spikes as large as 1700 volts have been measured in home wiring. Residential areas experience more large spike transients than commercial areas. The line filters in your power supply will protect the unit from some high-voltage transients, but occasionally a spike overcomes the power supply protection and gets to the logic circuitry. The usual result is erased automatic programs or a strange mode of operation, but if the spike is too large, sensitive circuit devices can be destroyed.

Your VCR power supply is normally not affected by the transients generated by on or off switching actions. These actions can produce a short-lived spike that is five times normal line voltage.

Spikes are not all generated outside the unit. When you begin to play a tape, for example, activating the tape drive mechanism produces a voltage spike inside your unit. Design engineers have placed capacitors in strategic locations on the circuit boards and in the tape drive mechanism electronics to carry the spikes harmlessly away to ground and prevent component damage. If any part of the spike reaches the circuit components, the devices are stressed and performance can become marginal.

Preventing Power Line Problems

If you live in an area where power outages or brownouts are common, where electrical storms occur when you aren't ready, or where your VCR occasionally shuts off and the clock display jumps to some strange hour, you need protection. You can prevent some power-line problems by using various forms of power-line conditioners, including isolators, the regulators, and filters.

Isolators provide protection from voltage and current surges and include transient suppressors, surge protectors, and other isolation devices. These devices maintain line voltage at a proper level even when the line supply is 25 percent over normal. Some surge protectors can filter out high-frequency spikes but cannot respond to slow, low-frequency transients. One form of surge protector is called a metal oxide varistor (MOV), a form of diode that will clamp the line voltage at a certain level, preventing over-voltage spikes from getting into your system. These devices are installed across the power-line wires leading into the unit. Isolators *cannot* provide protection against brownout or complete loss of electrical power.

Regulators act to maintain the line voltage within prescribed limits. They are essential if line voltage varies more than 10 percent at the unit, but they don't provide protection against voltage spikes and blackout.

Filters remove noise from the input power line. They short EMI/RFI signals to ground and remove high-frequency components of the signals from the low-frequency 60 Hz power line. Power-line filters work best when they are located immediately next to or at the front end of the power supply. Filters don't stop spikes, nor are they effective during low or high voltage conditions. When you select a power line conditioner, consider these factors.

- Speed of response in handling voltage spikes.
- Ability to filter out high-frequency noise.
- Ability to handle repeated transients.
- Amount of line power it can handle.
- Range of input voltages it will handle in producing clean power out.
- Multiple outlets to handle several devices.

The amount of power protection you provide is up to you. Many VCR users are able to get along quite well with unprotected systems. Others prefer to operate their systems knowing that the VCR is protected against unseen environmental upheavals.

CORROSION

The metal connector pins on cables, circuit boards, chip pins and other component leads are subject to corrosion, a chemical change in which the metal plating of the pins and sockets is gradually eaten away. Corrosion can damage not only the electronic connections in the VCR, but those connections in the mechanical section as well. Tape guides, solenoid plungers, levers and bearings are all subject to corrosion. If the smooth tape guide becomes corroded the oxide will be scratched off the tape surface. This corrosive buildup on the guide can permanently damage each tape operated in the machine. There are three types of corrosion that can affect the VCR system.

- Direct oxidation by chemicals.
- Atmospheric corrosion.
- Galvanic electrical corrosion.

Direct Oxidation

Direct oxidation is a form of chemical corrosion. A film of oxide forms on the metal surface and reduces the pin's contact with the socket or connection. At high temperatures the oxidation process accelerates. The metal is slowly worn away as the electrical contact surface is converted to an oxide and the oxide crumbles. It is this same oxidation that builds up on the tape guide surfaces and scratches the tape.

Atmospheric Corrosion

Chemicals in the air attack the metals in the VCR circuitry and metal chassis parts, causing pitting and buildup of rust. In the early stages of this corrosion, sulfur compounds in the atmosphere are converted to tiny droplets of sulfuric acid that lie on the surface of the connector pins. This acid eats away the metal causing pits to form.

When atmospheric corrosion is just forming, the plug contacts and metal parts can be wiped clean, restoring the metallic brightness. But if the sulfuric acid is allowed to remain, long exposure converts the acid to a sulfate layer that can no longer be wiped away.

The effect is reduced electrical contact between the pins and their sockets. A layer of discolored rust that prevents any contact between the pins and their sockets causes an open circuit and can be located. It's the in-between stage, when the "almost-open" condition exists, that produces those hard-to-find intermittent failures.

In addition, you should be especially careful if you live near the ocean. The presence of salt spray or increased levels of chlorides can cause severe pitting of some metals.

Galvanic Corrosion

In galvanic corrosion, an electrolyte such as salt in moisture penetrates between the metal plating and the underlying base metal through a tiny crack in the metal plating of a pin or connector.

A small galvanic battery forms, and a tiny electric current flows between the two dissimilar metals. The plating surface becomes scaly and rough as the plating is slowly eroded away. The corrosive action is concentrated on the underlying metal exposed at the breaks in the scale since this is where the galvanic battery exists.

The effect is the same as for the other forms of corrosion. Electrical contact between the pin and socket decreases, causing intermittent problems, until the scale is so complete that the electrical circuit is broken and the signals are blocked entirely.

You can actually start corrosive action by handling the connectors and boards improperly. The wrong way to handle printed circuit boards is demonstrated in Fig. 3.5.

Fig. 3.5. Handling a printed circuit board the wrong way can cause corrosion to occur.

Never touch contacts with your fingers. The oils on your fingers contain enough sodium chloride to begin oxidation action on those pins.

Corrosion Prevention

While metal storage sheds and cars can be spray painted to prevent rust (oxidation), this cannot be done on circuit boards, connectors and metal chassis parts. The best preventive action is cleaning and proper lubrication. By keeping the electrical contacts, tape guides and chassis parts clean, you can deter oxidation buildup and prevent the occurrence of intermittent operation and tape path trouble.

You can clean the pins on some plugs by reseating them periodically. Turn off all the power and carefully remove and push these devices back down into their sockets. This action will clean the pin surface and restore (or ensure) good electrical contact.

> **CAUTION:**
> Always turn off the power and touch a grounded surface before touching anything inside the VCR.

Oxidation of some contacts can be cleaned with emery cloth, a soft rubber eraser, a solvent wipe, or a contact cleaner spray.

> **CAUTION:**
> When rubbing to clean contacts, always rub along the pin (lengthwise). Rubbing lengthwise on the pins prevents accidentally pulling a pin contact up off the board.

If you use emery cloth, be careful not to grind away the metal plating itself. If you use a rubber eraser, keep eraser dust away from the unit.

Solvent wipes are found in the cleaning kits sold by many electronic supply companies. These wipes clean and lubricate the contact surface with a film that helps to seal out atmospheric corrosion without interfering with signal flow. Most solvent wipes are individually wrapped in small packages much like hand towelettes.

Spraying the pins with a contact cleaner spray (also available at most electronic parts stores) is an effective corrosion preventive. Contact cleaner wipes and spray are the best methods for removing oxidation.

There is a trade-off between preventing corrosion and preventing electrostatic discharge, because corrosive action is reduced with a reduction in the relative humidity, but ESD increases.

> **CAUTION:**
> Select a contact spray cleaner that will not harm plastics and that contains no lubricants. Lubricants contained in spray cleaners may get into the chassis and cause belts and tires to slip. If the spray is not properly controlled, the plastics in the chassis or the cabinet may also be damaged.

Electronics manufacturers are aware of the effects of corrosion, and most connectors are made of a combination of metals that resist corrosion but are good conductors of electrical signals. You can choose the type of connectors to use for your cables. You can buy cables and connectors with tin alloy plating on the pins or with a thin gold plating. Although you will pay more for the gold-plated connectors, they can be worth the price because they provide superior contact reliability. While gold-plated contacts don't wear out as tin alloy surfaces do, even the tin surfaces take a long time to wear away. A sound, consistent cleaning program can really help.

There is one final note on the subject of corrosion. High temperatures will increase the corrosive action in the VCR system, so keep your unit tuned up and running cool.

MAGNETIC FIELDS

The effects of magnetism are especially important in tapes and tape drive mechanisms, since these two parts of the VCR are based on magnetic principles.

Each cassette tape is coated with a magnetic oxide that has millions of tiny pole magnets randomly positioned on its surface. As the tape surface passes over the video head, a magnetic force is induced in the head by the signal current flowing through the head coil. This current causes the pole magnets on the tape surface to line up according to the video information being converted to voltage pulses in the head. This is "good" magnetism.

The voltages used in monitors and television receivers produce strong magnetic fields. These can be bad. If you accidentally place one of your tapes in the field, the tiny pole magnets on your tape's surface can change their alignment. Then when the VCR tries to read the tape, the head may misunderstand or misinterpret the information on the tape. You get garbage on the screen and in the speaker.

Magnetic flux can be caused by the presence of a high (115 V) voltage in monitors and televisions. A color television produces the strongest magnetic flux, but high voltage areas of monitors, telephones, stereo speakers, ballast in fluorescent lights, and even power strips can be sources of offensive flux and can cause a loss of recorded audio and video. The strength of the flux field depends on the strength of the voltage, which can fluctuate depending on the amount of power being required by the equipment.

The moral is: keep your tapes and hookup cables away from power sources and magnetic fields.

PREVENTIVE MAINTENANCE FOR VIDEO TAPES

Two valuable components in any VCR are the video tapes and the VCR itself. All Beta and VHS VCR systems use ½-inch video tape as the storage medium. Since tapes and tape drive mechanisms are such critical components in unit systems, it makes sense to do all you can to protect and maintain them.

The video head in the VCR rides on the surface of the most vulnerable part of the storage system, the cassette tape. Video tapes are made of a plastic base and coated with a magnetic iron (ferric) oxide.

Now tapes are pretty sturdy, but they are sensitive to magnetic and electrical fields, high temperature, low temperature, pressure, stretching, and dust. Dust and little airborne fibers are particularly bad. With the VCR's video head riding on the surface of the tape, any tiny piece of "junk" lying on your tape looks like a huge boulder. A piece of your own hair is about 40 microns (.0015748 inch) thick. A hair is a huge obstruction to the video head. Even dust and fingerprints on the tape surface cause obstacles to the even movement of the tape under the head. To protect the tape medium and the head, each tape has a door on the front of the cassette housing that shuts out dirt and fingerprints. The video heads inside the VCR rotate at 1800 rpm, and dust and other particles that may have slipped inside the door or settled on the tape are picked up by the tape guides and rotating video heads. A buildup of these particles can cause the tape to travel incorrectly along its intended path or prevent the heads from contacting the tape surface.

Tobacco smoke is harmful to tapes and tape drive mechanisms. The tars and nicotine that filter up into the air from the ends of cigarettes and cigars can settle on your VCR and form a gummy ash on any exposed surface including tapes and tape drive mechanisms. This material gums up the drive, eats into the video head, and scratches the surface of your tapes. The effect is similar to taking a metal file to your favorite record album. Avoid smoking or allowing smoking in your VCR area. If you can't do this, clean the system more often.

Tapes are further protected by storage boxes, or jackets. Use them. Don't let tapes lie around outside the sleeve inviting trouble from dust and dirt.

Not all tapes are created equal. Some tapes are manufactured to better standards, with thicker magnetic oxide coatings, and better binding materials. Naturally, these tapes are more expensive. Less-expensive tapes have thinner oxide coatings and shed their oxide layers easily. Compare tape specifications before you buy.

Depending on the quality of the tapes, the cleanliness of the VCR area, and the condition of the tape drive mechanism, tape life could be as short as a week or as long as many years. Assuming the quality, cleanliness, and condition factors are favorable, estimated tape life is based on actual passes through the unit while the video heads are in contact with the tape surface, rather than on total time of existence.

As the heads ride on the tape's oxide surface, they cause tiny bits of oxide to rub off. Some of these loose oxide particles are caught and stick to the heads. Gradually an oxide layer builds up. This oxide layer has two effects on system operation. First, it makes the heads less sensitive to recording and playing back signals. Second, it causes an abrasive action on the tape surface.

As the oxide layer builds up, it becomes ragged. This roughness scratches even more oxide off the tape, until the oxide layer on the tape surface is too thin to support information storage. When oxide is missing from the surface, "dropouts," or spots occur where data can no longer be stored. The tape fails to record or play back properly, and it becomes useless.

Keeping this oxide layer from building on the video heads will help extend the life of your tapes and machine. The better tapes are less likely to easily give off oxide particles, so the heads stay cleaner longer, and the tapes and heads last longer.

Toward Longer Tape Life

Here is a summary of what you can do to help extend tape life.

1. Buy name-brand tapes. Avoid "bargain" tapes. The $14 tape should last twice as long as the $7 tape.
2. Never touch the tape surface.
3. Never open the tape door on the cartridge (except during troubleshooting).
4. Store tape cassettes in their protective jackets.
5. Never force a tape into or out of a VCR. Place the tape in the unit carefully. Never slam the tape cassette holder down into the unit.
6. Store tapes in a cool, clean and dry place.
7. Don't lay tapes in the sun. They warp just like stereo audio cassettes and computer floppy disks.
8. Never allow smoking near the tapes or VCR. Smoking causes tars to settle on the tape surface (and inside your VCR), gumming up the works.
9. Never set tapes down on or near monitors, televisions or speakers. The magnetic fields can erase programs.
10. Avoid placing tapes near vacuum cleaners or large motors. Even freezers and refrigerators have compressor motors that can erase programs on your tapes.
11. Store tapes vertically. Storing tapes

horizontally can cause the tape edge to bend and to bind in the cartridge. This causes damage to the edge or tape surface.

12. Don't put tapes through airport x-ray machines. Hand them to the security guard for inspection, and have them bypass the x-ray inspection process.

PREVENTIVE MAINTENANCE FOR VCRS

What kind of preventive maintenance is there for VCR mechanisms? How can you test and maintain your own VCR?

Many VCR manufacturers state that there isn't any preventive maintenance that could be done by a novice. Then they describe head cleaning as the only routine maintenance that a customer should do and then only with a head cleaning cassette. Other manufacturers do not recommend cleaning tapes and suggest cleaning be done only by trained service technicians.

Why is this so? The manufacturers have great concern for the critical mechanical and electronic alignments inside the machine. They are concerned that untrained hands will accidentally damage this complex electromechanical device.

Many adjustments should only be made with high precision alignment jigs and electronic test equipment. There are even some VCR alignments that can only be performed at the manufacturer.

An incorrect half turn of alignment screws in the tape path, such as those shown in Fig. 3.6, could require hours of realignment work for a trained technician. This could be very expensive to you.

CAUTION:
Make no alignments or adjustments on any tape path or other mechanical or electronic components without being absolutely sure of what you are doing. Follow only the prescribed procedures in the machine service literature.

Fig. 3.6. Turning the head Height Adjust or V-Block Position screws could severely affect the alignment of the mechanics of the VCR.

The repair business is big business. The less preventive maintenance you do or have done, the sooner your VCR will start giving problems and the more work you will provide for repair companies. Here are some facts.

Cleaning the Heads

Heads need cleaning to remove the oxides building up on the leading edge of the head. Head cleaning is preventive maintenance that you can do using any of the various head cleaning cassettes that are available. The "wet" kind works with a cleaning solvent.

Some head cleaners are abrasive and can damage the head if they are used for too long. If you buy this type of cleaner, you must use it just long enough to remove the oxide but not long enough to damage the heads. Recently, nonabrasive head cleaners have begun to be marketed. One head-cleaning kit uses fabric-tape dampened with cleaning solvent. With this kit, you sprinkle cleaning solvent on the fabric and then insert the cassette into the VCR for spinning action head cleaning.

Since any cleaning cassette works by rubbing action and chemical reaction between the tape fabric and the video head, there is potential for abrasion. So be careful not to leave the cassette in the VCR for too long. Carefully follow the instructions that come with the cleaning kit.

Video heads can also be cleaned with certain types of alcohol or freon TF and lint-free materials like the ones shown in Fig. 3.7.

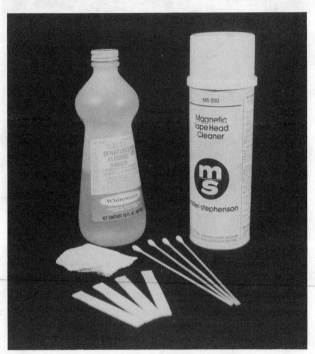

Fig. 3.7. A chamois, some swabs, and the solution or spray used in cleaning a VCR.

Special cleaning material such as cellular-foam swabs, chamois leather cloth, or a piece of special lint free cloth are good materials to use for manual head cleaning. Regular cotton swabs are dangerous because the cotton fibers can catch or pull away and lie in the tape path or on the head. They can also catch on the ferrite chip in the middle of the brittle head, loosen it from its mounting, or actually break the core and ruin the head.

Special cotton swabs can be used to clean the tape path. These have tightly wound cotton at one end and are suitable only for the tape path—not the heads. Denatured alcohol or methanol can be used as a cleaning solvent as can typewriter cleaner or trichloroethane. The solvent must not leave a residue when it evaporates, so most other alcohol solvents should be avoided. Always have plenty of ventilation and make sure the solvent has evaporated before you operate the machine.

The frequency of cleaning depends on how much the VCR is used and the type of cassettes that are used. A quality tape is good for several hundred passes through the machine. Remember both record and playback time must be considered.

Almost all manufacturers recommend that video heads and the entire tape path be cleaned every 500 hours of operation. This includes record and playback hours. If rented movies are played in your unit on a regular basis, more frequent cleaning is a good idea. You never know the condition of the previous user's machine.

Bargain tapes are good for about one-tenth the life of a name brand tape. This means that instead of 200 hours of record/playback time, you might get 20 hours or less before the head gets caked with oxide or the tape surface becomes too worn to record or play back.

Keeping the cassette door closed on the VCR unless you are inserting or removing a tape will help prevent dust and dirt from getting inside. It also prevents unwelcome visitors (insects and even mice) from climbing into the unit. Believe it or not this does happen!

Preventive maintenance also involves regular cleaning and lubrication of mechanical parts as well as replacement of worn or stretched rubber tires, belts and torque limiter assemblies.

Cleaning Cassettes

Before you groom your VCR using a cleaning cassette, there are some things you should know. Generally speaking there are two types of cassette cleaning tapes on the market today. One is known as a "dry" cleaning tape, the other as a "wet" cleaning tape. The wet cleaning tape is a chamois material with drops of cleaning solution placed on it. In some cases the material is much thicker than normal video tape.

Once the wet cleaning tape has been moistened as directed by the instructions, it is placed in the video recorder and the VCR is switched to the play or record mode. The moistened chamois is designed to travel in the tape path and, as it is pulled against the video and audio heads, the heads are wiped clean. It takes less than a minute to wipe off any contaminants that may have accumulated on the video head and in the tape path.

The dry video cassette cleaning tapes are

used the same way, but do not require a cleaning solution. Dry tapes come in two basic forms. One of these uses a fibrous material to wipe over the heads with a gentle rubbing action. The other type looks like normal video tape, but is more abrasive. The abrasive material actually scrapes off contamination.

Video cleaning cassette tapes have occasionally damaged video tape recorders. The cleaning cassettes using a chamois material have been known to unscrew or break tape guides. They have also lodged under the video heads and broken the assembly. A contributor to this problem is the thickness of the chamois in the tape path. At least one manufacturer of "wet" tapes has recently reduced the thickness of the cleaning tape (Fig. 3.8) to decrease the potential for damage.

Fig. 3.8. An improved wet tape cleaning kit with thinner cleaning tape.

If the instructions for the wet cleaners are not carefully followed, it's possible for the cleaning chamois to become saturated with foreign particles and cause machine damage. This can occur if the cleaning cassette is used too many times. If this happens, the cleaning tape can deposit contamination in the tape path.

Occasionally a video head becomes so clogged by excessive use that a wet cleaning tape may not completely remove the contamination, and the video head assembly will seem defective, even after several cleaning attempts. You may need to use a dry tape.

The dry cassette cleaning tapes which look like normal video tape are very abrasive. The manufacturers of these tapes state that operating a cleaning cassette in your machine for thirty seconds produces the equivalent head wear of running a normal tape through your machine for six to twelve hours. These dry video tape cleaners are not as likely to damage tape guides and rollers because the tape is the same thickness as normal video tape. Nor are we aware of any of these tapes breaking a video head. However, if used too often, this type cleaning system will cause excessive wear on the video heads. The manufacturers recommend that you only use them when you actually experience symptoms of a clogged head (snow in the screen display or no video).

There have been instances in which a wet cleaner and repeated hand cleaning did not remove buildup from the video heads. Then an abrasive dry head cleaner was used for a few seconds, and it did the job.

The other type of dry cleaning tapes use fibers in the tape that are not as abrasive.

The California State Electronics Association is currently asking servicers to report any troubles caused by video tape cleaning cassettes. This information will be passed to an official consumer protection agency in California. It's likely that a consumer advisory may be issued for the use of cleaning cassettes in California. These results will benefit VCR users throughout the world!

If you choose to use a cleaning tape, carefully follow the instructions provided. This is the key to getting cleaning cassettes to work properly for you. If excessive snow or no video persists after using these tapes, have your unit checked by a trained servicer. The failure may be in the electronic circuitry or in the video heads themselves.

A good quality head cleaning cassette is worth owning. Nothing is more frustrating than to be enjoying a good movie with your friends and to suddenly lose video because of a clogged head. One head cleaning cassette can make an evening. But, if your recorder is still under warranty, check with the manufacturer first before you use a cleaning cassette.

Follow these hints to clean video heads with a wet cleaning cassette.

1. Read and understand directions thoroughly.

2. Turn on the VCR power.
3. Dampen the cleaning cassette with the solvent that comes with the kit.
4. Insert the cleaning tape in the unit.
5. Place the unit in PLAY.
6. After the instructed time stop the VCR and remove the cassette.
7. Turn off the VCR.
8. Let the tape path dry completely before operating the system.

To clean video heads using a dry cleaning cassette, follow steps 1, 2, 4, 5, 6 and 7 in the preceding list.

Cleaning the Video Heads Manually

To clean the video heads and tape path manually, first be sure you have these supplies.

- Flat head screwdriver.
- Phillips head screwdriver.
- Adequate lighting.
- Tray to hold loose screws.
- Cleaning material.
- Cleaning solution.

When you have assembled the necessary equipment, turn off the power and unplug the VCR from the AC socket. Remove the top cover of the VCR using the procedures found in the service manual. Also remove all internal shields from around the video heads and tape path to provide more working room inside the unit. Be sure that ground wires removed from the shields are not touching the unit.

Now, reconnect the power and turn on the VCR. Place a tape in the unit, press PLAY, and observe where the tape comes in contact with the tape guides, capstan shaft, audio and erase heads and the rotating video head drum as shown in Fig. 3.9. Also note how the tape is guided along a ledge on the lower stationary portion of the head drum.

Stop the unit, eject the tape, turn off the VCR and pull the power cord plug out of the electrical socket. Moisten the cleaning swab material with cleaning solution and clean all of the guides where

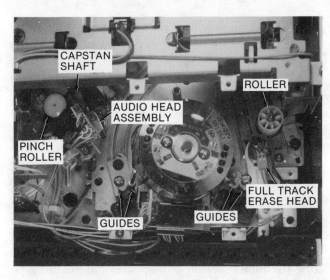

Fig. 3.9. The tape path in a typical VCR.

the tape makes contact. Before you clean the video head, read this.

> **CAUTION:**
> APPLY ONLY THE SLIGHTEST PRESSURE POSSIBLE TO THE VIDEO HEADS. MOVE THE CLEANING INSTRUMENT IN THE SAME PLANE AS TAPE TRAVEL. HEADS ARE FRAGILE, AND CLEANING IN AN UP AND DOWN MOTION CAN BREAK THE HEADS.

Fig. 3.10 shows the upper drum (upper cylinder), a video head, the lower drum, and the lower tape guide ledge.

Hold the rotating portion of the video head at the very top of the drum to stop it from moving, and with a gentle side-to-side stroking action, clean the video heads and the upper drum assembly. Hold the head drum only at the *very* top. Carefully rotate the drum to clean the rest of the upper drum. While cleaning the lower (stationary) part of the drum, carefully rotate the upper video head drum keeping the heads away from the cleaning instrument. This prevents accidental contact with the fragile heads. Pay special attention to the lower tape guide ledge around the stationary portion of the drum. Any buildup here will cause tape path error.

Next, clean the audio control track and erase heads. When you do, move the cleaning swabs in

Fig. 3.10. The upper drum and the lower drum with its tape guide ledge are clearly visible in this photograph. A video head can be seen between the upper and lower drum assemblies.

the same plane as the direction of tape travel. A forward and back movement is good.

Clean the capstan shaft and rubber pinch roller. Do this last because the capstan frequently has the most debris on it and the pinch roller will get your cleaning swabs all black. If you did this first, you might add contaminants to the heads or guides.

Now, throw away the cleaning instrument. If it were used again, it could cause future contamination of the heads. Don't be afraid to use more than one cleaning swab on the tape path each time you clean, this is no time to be cheap. If a dirty cleaning stick is dipped into a bottle of cleaning solution, the dirt from the stick will come off in the solution and eventually find its way back into the system again.

Let surfaces dry completely before you reassemble the unit. When the heads and tape path are dry, reinstall the shields, attach any grounds and replace the cover. Attach the AC cord, make a test recording, and play back to check the audio and picture quality. (Note: if you have any problems refer to the specific troubleshooting chapter in this book.)

Video Head Cleaning Interval

Cleaning your tape path is like changing the oil in your car. You do it when you feel you've driven enough miles or when the oil looks dirty. Most VCR manufacturers recommend cleaning heads every 500 hours. Some repair centers suggest cleaning every six months; others suggest you don't clean the heads until the TV picture has a lot of snow in it or the audio sounds bad. If you live in an area that gets a lot of smog, you may want to clean the heads more often. In any case, it won't hurt to clean at least annually.

Mechanical Preventive Maintenance is not needed as frequently as head cleanings. Belts, rubber tires and torque–limiting clutch assemblies are changed or cleaned every 1500 hours of operation. Replacement recommendations vary among manufacturers but most are in the 1000 to 2000 hour range. Rubber will stretch and crack with time, so even if you don't use the VCR a lot, you should consider replacing the rubber parts every 18 to 24 months. Unfortunately, failures always seem to occur at the wrong times.

One customer stated that she had 20 children over for her nine-year-old's birthday party. She rented movies as the main entertainment because it was raining heavily. Guess what happened? A two-year-old belt in the VCR broke in the middle of the video movie. Preventive maintenance is designed to prevent such experiences, but it only helps if you do it regularly. The Appendix has a suggested preventive maintenance schedule for VCRs.

Preventive Maintenance for the Tape Drive Mechanism

This procedure is not recommended for the person who is not mechanically inclined or who doesn't like getting a little dirty. If this is you, refer the job to a trained servicer. If you decide to try anyway, you'll need a service manual which you can buy from the manufacturer. In addition, there are some important things to know before you start.

Don't get grease or lubricant on the belts, pulleys or rubber tires because oil or grease can cause these parts to slip and lose traction. Loss of traction prevents proper operation. And use only small amounts of lubricant. Half a drop of oil goes a long way in a VCR. Too much lubrication is bad because oil can migrate to and damage other parts of the machine.

Use lightweight high-grade machine oil or an

equivalent. Transmission fluid is an excellent substitute because it is light, clean, and does not gum up with time like many oils. But it can also migrate easily, so use it sparingly.

Use only the greases that are recommended by the manufacturer. Some units require different types of grease at different points. The wrong grease (such as standard petroleum grease) will react with the nylon, plastics and metals and cause the moving parts to stick. Petroleum greases gum up faster than most synthetic VCR greases, but even some synthetic greases cause the machine to malfunction much sooner than expected. Fortunately, new greases have recently been developed that don't thicken.

Never disassemble anything in the tape path including the blocks and supports that the guides travel and seat in. Alignment of these areas is critical and in some cases cannot be done except at the factory (and most of these factories are in Japan).

Thoroughly remove all the old grease or oil before applying the new. And don't forget the shaft and collars in which the pulleys are mounted. Remove the pulley shaft and wipe the shaft and collars with a cleaning solvent on a surgical cotton swab, like the one you used on the tape path. Remember to clean and lube the pulley shaft before reinstalling it.

Never force anything. If it won't move, it's either incorrectly positioned or it isn't supposed to move. Never use spray cleaners or spray lubricants in the chassis. Chances are good the spray will cause more problems than it will prevent.

Check all belts and rubber tires for wear or age before removing them. Stretch the belts as shown in Fig. 3.11, and pull the rubber slightly away from the tire rim to check for cracks as is being done in Fig. 3.12.

If the belts or tires appear to be dry, cracked, worn or stretched they must be replaced. It's important to have replacement parts at your side BEFORE you remove the old ones. This way you'll remember which belt went where. Also, if the parts are not readily available, your unit is not out of commission for an extended period of time.

Read the service disassembly instructions and carefully note any warnings. Do only those things that you are confident you can handle. If you are unsure about a procedure don't do it. You may end

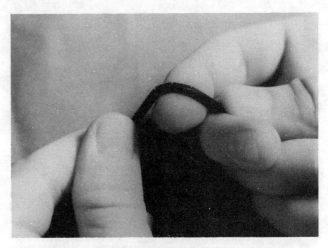

Fig. 3.11. Stretching and bending the belt out can reveal hidden cracks.

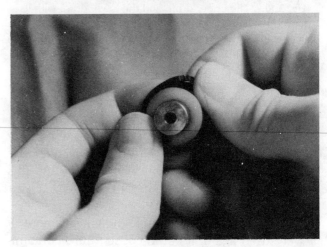

Fig. 3.12. Rotate the tire away from the rim and look carefully for signs of cracking.

up paying a servicer to untangle what you have done.

Take notes or use some other method to remember where each of the screws, washers, etc., go in the unit. The sizes, lengths and threads are often different for each area in the chassis. An ice cube tray or a muffin tin are both good small parts holders. The screws from each dismantling step can be organized in these holders to help you during reassembly.

As you work, watch for tiny washers beneath the tires and pulleys which often stick to the item being removed and disappear onto the floor. These washers, often made out of graphite, are essential to the smooth operation of the unit.

With these observations in mind, you're now

ready to disassemble your own unit. Remember to work patiently and carefully.

Fig. 3.13. Most printed circuit boards swing out for easy access during troubleshooting and repair.

First, disconnect all power and antenna connections. Place the unit on a flat, clean working surface. Now, carefully reread the service literature, paying special attention to those sections on which you are going to be working. Remove the top and bottom covers of the VCR as directed in the service data. Remove the cassette tray assembly, and open the necessary circuit boards on the top and bottom. Instructions for these procedures are in most service manuals.

Each VCR has a "service position" for opening the boards and working on the bottom part of the chassis. This usually requires standing the unit on one of its sides. On most units the circuit boards are on hinges that swing open for easy access as shown in Fig. 3.13. These boards don't require complete removal.

Occasionally, manufacturers identify the disassembly screws in the circuit boards and cassette tray assembly with special colors, red is common. If your machine is one of these units, your job will be much easier.

Remove all of the belts and clean the pulleys the belts ride in with a special cotton swab. These swabs are ideal because the cotton is tightly wound on the stick and doesn't fray easily. Fig. 3.14 shows the belts and pulleys to be cleaned on one VCR.

Fig. 3.14. Check and clean the belts and pulleys shown here.

Carefully remove the rubber tires on the top side of the unit. Be sure to collect those tiny washers! Reel tables, a torque limiter assembly, rubber tires, and belts are shown in Fig. 3.15. The reel table, lower left, has a tire that is shedding rubber dust in the slide lever to its left.

Pull the under side pulleys (attached to the shafts associated with the tires you just removed from the top of the unit) out of their collars. Any more washers?

With a swab soaked in solvent, carefully clean the metal collar (often brass colored) in which the pulley shaft spins. Fig. 3.16 shows a pulley shaft after removal. The finger points to old grease buildup which should be wiped off.

After the solvent has dried, apply a very small amount of lubrication to the pulley shaft, replace the shaft and new tire.

When you replace the tire, make sure it spins freely. If the tire is placed on the shaft without a little clearance, it will friction bind after the unit warms up and the collar swells. If the tire snaps on,

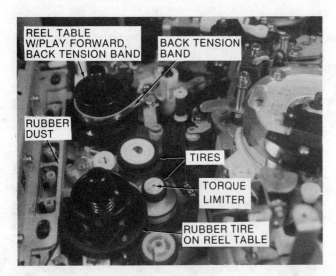

Fig. 3.15. Rubber dust is quite visible on the slide lever to the left.

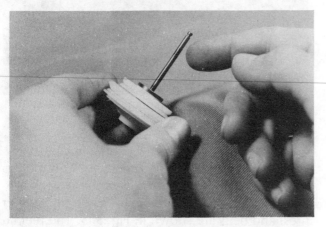

Fig. 3.16. Carefully remove the grease buildup and lubricate the pulley shaft before reassembling.

as many do, there is no clearance adjustment. Don't forget the washers!

Replace the worn torque limiter assemblies. If these are on shafts, clean and lubricate the new shaft just as you did the shafts of the rubber tire pulley.

Install new belts on the unit. Remove any twists in the belts. A twist can cause a belt to work out of the pulley groove. Replace the bottom circuit boards and return the unit to its normal operating position.

Now you have a decision to make. Part of the mechanical Preventive Maintenance involves checking and adjusting tensions and torques in the cassette drive mechanism. These adjustments re-

quire expensive gauges like those shown in Fig. 3.17 so to perform these checks you will need to borrow or purchase the gauges. If this is inconvenient or not desirable, have a service shop check and align the tension and torques for you.

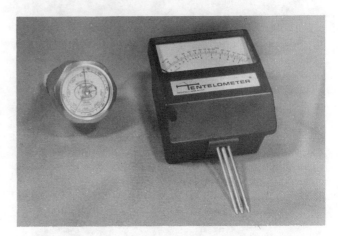

Fig. 3.17. Typical gauges used in tension and torque adjustment.

The following section on adjusting torques and tension is essential to extending the life of video tapes. If you aren't going to perform this adjustment, you can skip to the checkout and reassembly instructions below. On the other hand, you might want to read this section so that you understand why tension and torques are so important.

Adjusting Tension and Torques

The video tape must wrap around the video head drum under precise tension. If the wrap is too tight, the heads and tape surface will wear out from excessive friction. If the wrap is too loose, the heads do not contact the tape properly and the picture is lost. Correctly setting this tension is called *play or forward holdback tension adjustment*.

Torques are similar to tension but they serve a different purpose. The torques are the amount of power the reel tables have as they rotate. For example, the tape is pulled through the VCR in fast forward or rewind by the torque of the engaged reel table. There is a play torque too. The tape is pulled back into the cassette in play or record by the take-up reel table that's on the right side facing from the front in a VHS unit (left side for Beta). The amount of pull this reel table has is known as *take-*

up or *play torque*. If this pull is too high, the tape may be stretched. If it is too low, the tape won't be pulled evenly into the cassette and may not be pulled at all. Tape will spill into the machine and be damaged. By the way, the VCR senses the take-up action, and if it senses no take-up torque and the tape is spilling, the unit will stop functioning.

When the VCR is in the STOP mode, brakes are applied to the reel tables to halt the tape movement. The amount of braking force is called *brake torque*. If the brakes aren't doing their job, the tape comes crashing to the end of the leader each time you FAST FORWARD or REWIND, increasing the possibility of the tape snapping. If the brake torques are too high, tapes will stretch each time the VCR is stopped. Proper brake torques are very important.

There are several ways to check and adjust tensions and torques. A typical device to measure tape tension is shown in Fig. 3.18.

Fig. 3.19. Measuring rewind torque using a special test instrument.

a spring, while other tensions and torques are derived from the belts, tires and torque limiter assemblies (felt clutches). Follow this series of steps for checkout and reassembly.

1. Replace the cassette tray assembly.
2. Connect the VCR to the antenna and TV.
3. Plug in the power cord and energize the machine.
4. Make a test recording and play it back. Use a new tape so you don't lose valuable material.
5. Observe the tape travel. Watch for damage to the tape as it moves through the machine. Watch the tape in both load and unload operations as well as fast forward and rewind functions.
6. If you notice incorrect operation, stop the unit and refer to the specific troubleshooting section in Chapter 4.
7. Turn off the VCR, disconnect the power cord and install the top and bottom covers.

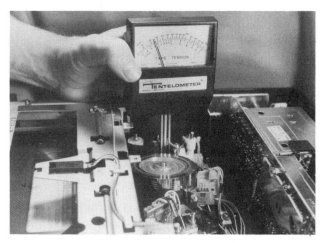

Fig. 3.18. A device measuring tape tension adjustment.

Fig. 3.19 shows how to measure rewind torque and supply reel table brake torque.

Consult your individual service manual for the method and types of gauges to use. Basically, however, place the unit in the mode to be tested, and measure the tension or torques with the appropriate gauge. Some units use direct drive reel tables in which the motor and reel tables are one part and there are no connecting belts or tires. These must be adjusted electronically. If tires and belts are used to drive the reel tables, play holdback tension is controlled by a felt brake band and

SUMMARY

This chapter has covered every aspect of routine preventive maintenance necessary to keep your VCR in peak operating condition. It discussed six major contributors to VCR failures: excessive temperature, dust buildup, noise interference, power line problems, corrosion, and magnetic fields. For each of these factors, this chapter presented one or more preventive countermeasures. You learned that video cassettes are to be protected from dust and dirt, discovered how tapes and video heads can be damaged, and most important, learned how to extend the life of tapes and tape drive systems.

CHAPTER 4

Specific Troubleshooting & Repair

Chapter 4 is a specific troubleshooting and repair guide focusing on a large variety of VCR failures. The section is divided into six parts.

- Audio problems.
- Video problems.
- Color problems.
- Power supply problems.
- Improper operation or no functions.
- Clock and timer problems.

Each fault can be associated with one of these areas. By letting your "fingers do the walking" through the Troubleshooting Index, you can quickly locate the page where your particular problem is addressed.

Part 1 covers audio-related problems including no sound at all, varying levels of loudness, unwanted audio still on the tape, weird-sounding audio, and static-filled sound.

Part 2 discusses all the symptoms related to video problems including no video, snowy pictures, intermittent video, strange video shapes, and bands of interference on the screen.

Color problems are addressed in Part 3. This section covers no color, and strange color.

Part 4 addresses all symptoms that can occur after you turn your machine on and discover power partially out, or totally gone. This includes the problems of no working functions, eject working only, and operational malfunctions.

In Part 5 you'll find the symptoms related to improper or inoperable functions such as sluggish rewind, mode functions that change after initiation, no play, fast forward, rewind, or record capability, and tape spilling into the machine.

Clock and timer problems are covered in Part 6. This section covers the situations where the timer loses 10 minutes every hour and where the machine won't auto-record using the timer.

TROUBLESHOOTING

VCR failures can be categorized as either electrical or mechanical. Each part of this chapter is subdivided into specific failures and provides symptom, troubleshooting steps, and possible cause. The step-by-step troubleshooting instructions quickly guide you to the electrical or mechanical cause for most VCR problems.

Many malfunctions are caused by contamination or foreign objects blocking the tape path or mechanical moving parts.

Isolating a failure to a mechanical or electrical breakdown can sometimes be challenging. Mechanical problems may cause the electronic circuits to act strangely and appear at fault. There is a close relationship between the electronic and mechanical functions in a VCR.

This chapter will help you identify and localize a fault to not only an electronic or mechanical failure but, in most cases, to the circuit or function in which the problem has occurred. Possible causes are listed to assist in determining the exact malfunction. The faults are categorized into functional areas. By matching the symptom that you discover with the closest description in the Troubleshooting Index, you will quickly be able to determine if the problem is one that you can (or want to) correct yourself.

As you progress through the troubleshooting charts, you can decide if you have the ability to service the unit yourself. Once you step past a certain point, most problems require service center action, or at least advanced troubleshooting techniques and test equipment. This book does not assume you have these skills. If you'd like to try the advanced methods, refer to Chapter 7 for guidance. Always observe good troubleshooting procedures. However, if you are not experienced in electronic servicing or don't have the right test equipment, you should refer the advanced troubleshooting to a qualified servicer. Nearly all component replacements require soldering skills. Understanding the chapters titled "Advanced Troubleshooting," "Magnetic Recording Theory," and "VCR Operating Theory" is necessary before attempting any advanced troubleshooting.

WHEN SHOULD YOU CALL THE REPAIR SHOP

The message is clear, if, during your repairs, you reach a point where you're unsure about your techniques or skills, STOP and refer the repair to a qualified technician. It's less expensive to pay up front for a repair that requires skills beyond those that you have, than to pay to undo a "mistake."

An article in the June 21-27, 1986 issue of TV Guide magazine described an Electronics Industries Association (EIA) study which found that 32 percent of all VCR failures were corrected by cleaning. Cleaning and belt replacement are the most common cures for VCR malfunctions. The next most common repair action (15 percent) was to the rewind mechanism; this was closely followed by video head replacement (11 percent). We've found that video head problems account for less than 6 percent of all failures in our business. Nevertheless, most VCR failures can be solved by a good cleaning and belt replacement.

Cleaning and simple mechanical adjustments/replacements are types of service you can do. This chapter is designed to help you diagnose and correct most failures, or decide whether to attempt the repair or refer it to a trained technician.

SPECIAL TERMS

In the troubleshooting charts there are some VCR terms you should be familiar with. These terms are described here.

E-E (Electronic-to-Electronic): the VCR video and audio output signals generated when the TV/VCR button on the unit is in the "VCR" position. In E-E, the VCR output signal to the TV is the same signal received by the VCR's tuner or line input jacks and processed by the unit.

Linear Track Audio: the audio signal recorded on the tape by the stationary audio head of all VCRs.

AFM (Audio Frequency Modulation): an option available for recording audio in High Fidelity stereo. In this method, heads mounted on the rotating video head drum assembly produce high quality audio playback.

USING SAMS PHOTOFACTS

For those problems that require service manual support, we recommend you obtain a Sams PHOTOFACT covering your machine. These manuals are available at most parts stores. Each

PHOTOFACT is a detailed troubleshooting manual that includes circuit voltages as measured, oscilloscope waveforms at key points in the circuitry, and circuit resistance measurements. Since most VCRs are designed and built in Japan, a form of "Japan-English" or "Japanenglish" writing is a characteristic of many manufacturers' documentation. Sams has done an excellent job of summarizing this documentation in their PHOTOFACTS. While the manufacturer's service manuals have more block diagrams, a theory of operation, expanded troubleshooting charts, and a lot of information, the printed language can be difficult to follow and understand. Combined with this VCR troubleshooting and repair guide and a Sams PHOTOFACT on your own machine, you are in optimum position to conduct your own advanced troubleshooting and repair.

> **NOTE:**
> The following troubleshooting techniques may require soldering. If you are uncomfortable with this, go as far as you can without soldering and then consult your local VCR service center.
>
> Desoldering or soldering on your VCR may void your warranty.

FAILURE/SYMPTOM INDEX
AUDIO

Page

Won't Record Audio. Video O.K. 52
Playback Volume Level Fluctuates 54
No Audio During Playback 55
Wow and Flutter 57
Audio From Previous Program Still Heard 58
Buzz Or Hum in Audio Playback 58
Low Audio Level in Playback of Self Recorded
   Tapes. Pre-recorded Tapes Playback With
   Normal Sound Level 59
Low Audio Level During Playback of Both
   Self and Pre-recorded Tapes 61

Audio Plays Back at Wrong Speed 62
No Audio 62
Static or Popping Sound in Hi-Fi Audio 63

VIDEO

Snow on Screen 64
Fine Horizontal Line Floats Through
   Picture 66
Picture Alternates Between Good Video
   and Snow 67
No Video 68
Vertical Images Bend or Tear Near Top of
   Picture 69
Snow Bands or Lines of Interference at Top
   or Bottom of Picture 70

COLOR

No Color on Newly Recorded Tapes.
   Pre-recorded O.K. 71
No Color in Record or Playback.
   Pre-recorded Tapes Have No Color 73
Bands of Color on Screen During Color
   Playback 75

POWER SUPPLY

No Functions. Clock May or May Not
   Have a Display 76
No Power (dead VCR) 77
No Modes Functional. Only Eject Works 78

IMPROPER OR NO FUNCTIONS

Only Eject Works 79
Shuts Down or Returns to Stop After a Few
   Seconds. Tape May Spill Into Unit 80
Won't Properly Rewind Tape 81
Rewind Stops Before End of Tape 82
Won't Rewind at All 83
No PLAY, FAST FORWARD, REWIND or
   RECORD 84
Tape Spills Into VCR 87
Slow or No FAST FORWARD 88
Mode Button Won't Engage on First Try 88

CLOCK/TIMER

Clock Loses 10 Minutes Each Hour 89
No or Intermittent Record Using Timer 90

AUDIO PROBLEMS

SYMPTOM: Won't record Audio. Video O.K. TV has Audio on normal TV.

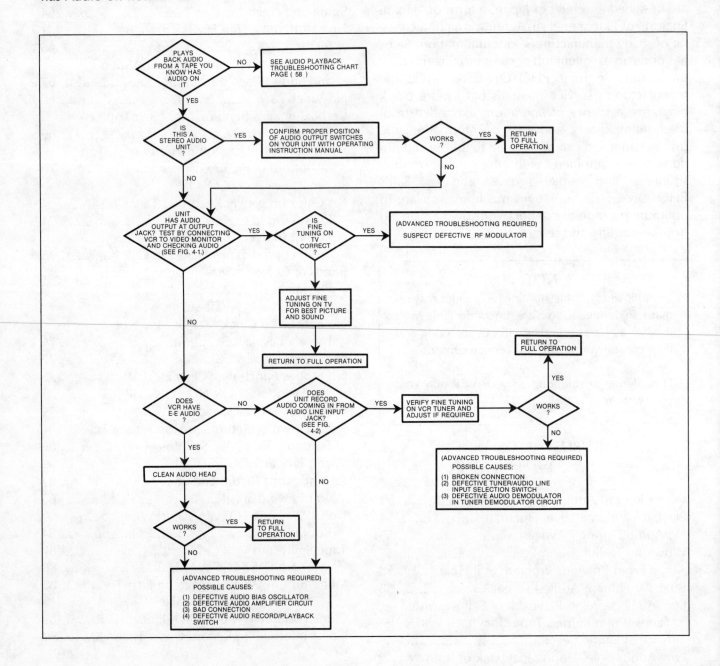

Specific Troubleshooting & Repair   53

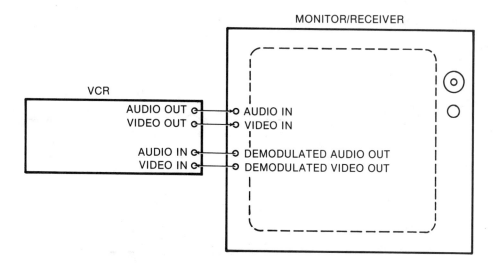

Fig. 4.1. Connect the VCR Audio Out to the monitor Audio In.

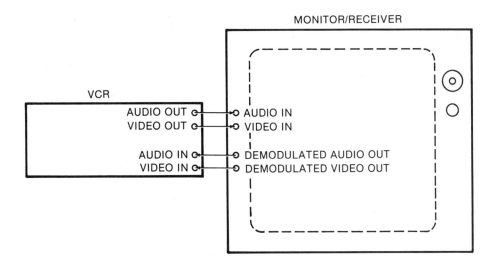

Fig. 4.2. Connect monitor Demodulated Audio Out to VCR Audio In.

SYMPTOM: Playback volume level fluctuates.

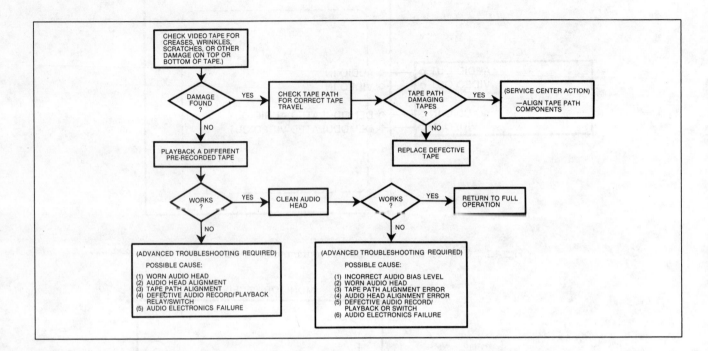

SYMPTOM: No Audio during playback. Video O.K.

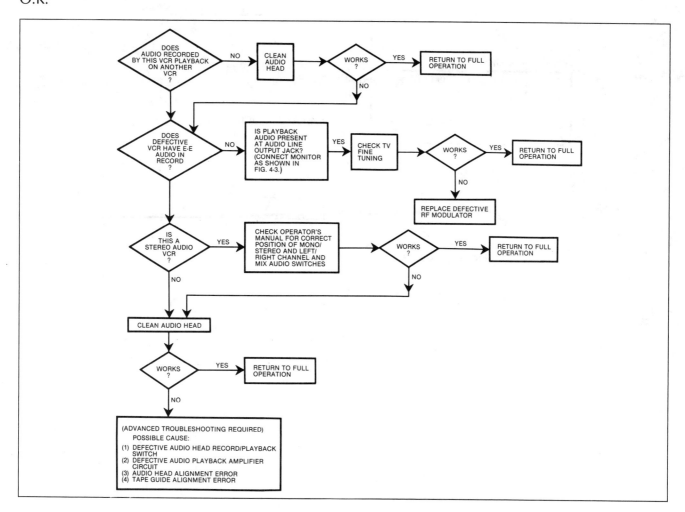

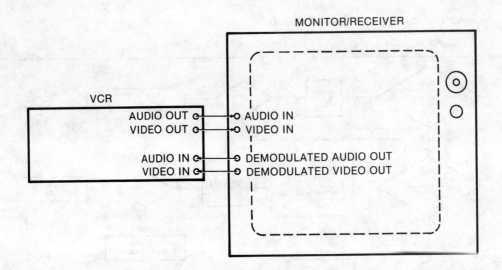

Fig. 4.3. Connect VCR Audio Out to monitor Audio In.

Specific Troubleshooting & Repair

SYMPTOM: Audio has Wow and Flutter. Sound varies in pitch (not volume) and is especially noticeable on sustained notes in music.

SCENARIO

You just made a recording of one of your favorite musicals. When it is played back the sound seems to vary in pitch at regular intervals. The problem is especially noticeable in the music portions of the recording. The pitch of the sustained tones seems to vary. The volume level doesn't change but the pitch changes. One description of the symptom might be to say the VCR has a bad case of vibrato.

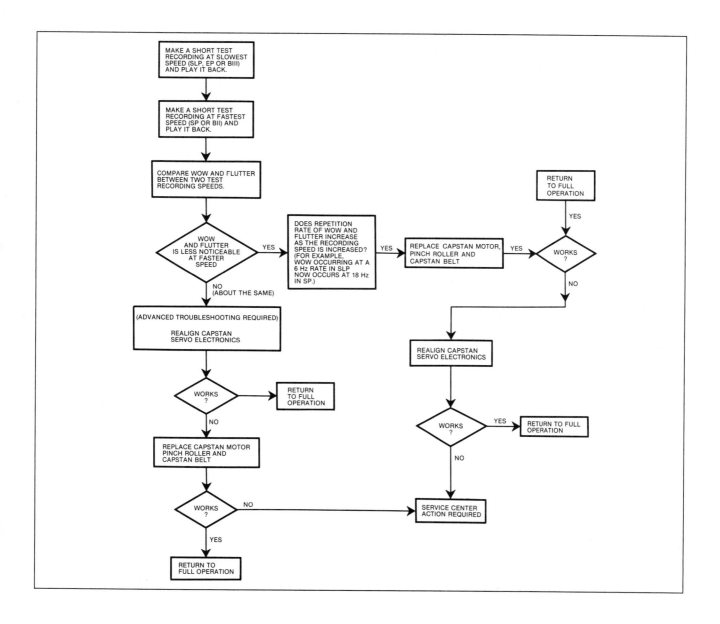

SYMPTOM: Audio from previous program still heard. Records Video O.K.

SCENARIO 1

You are anxious to record a program on TV. You select a video tape you have recorded on before and quickly start recording. When the program is over, the tape is rewound and you start to play back the new recording. You can hear the sound of the new recording OK but mixed with the new sound is some sound from the program you previously recorded on that tape.

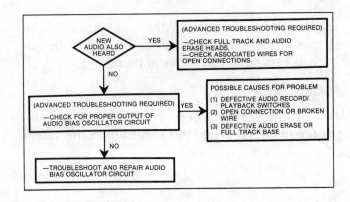

SYMPTOM: Buzz or hum in Audio playback.

SCENARIO 2

Same as Scenario 1 except that during playback, you hear only the previous audio.

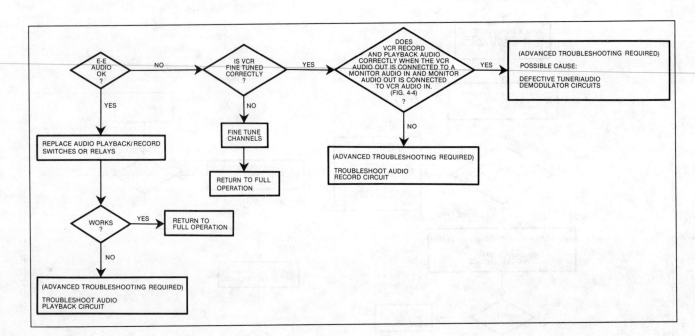

Specific Troubleshooting & Repair 59

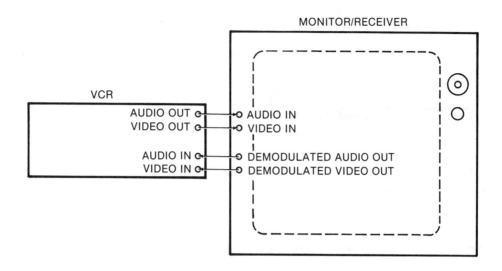

Fig. 4.4. Connect VCR Audio Out to monitor Audio In and monitor Demodulated Audio Out to VCR Audio In.

SYMPTOM: Low Audio level in playback of self-recorded tapes. Pre-recorded tapes play back with normal sound level.

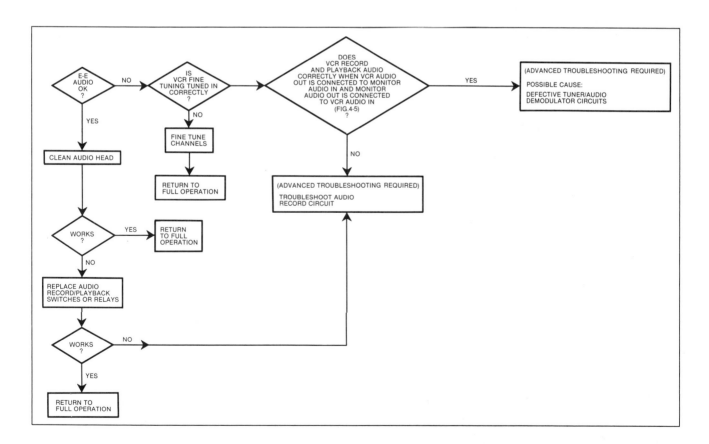

60 VCR TROUBLESHOOTING & REPAIR GUIDE

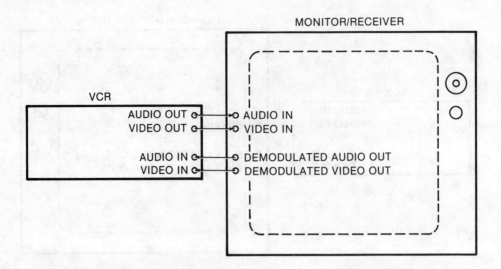

Fig. 4.5. Connect VCR Audio Out to monitor Audio In and monitor Demodulated Audio Out to VCR Audio In.

SYMPTOM: Low Audio level during playback of both self- and pre-recorded tapes.

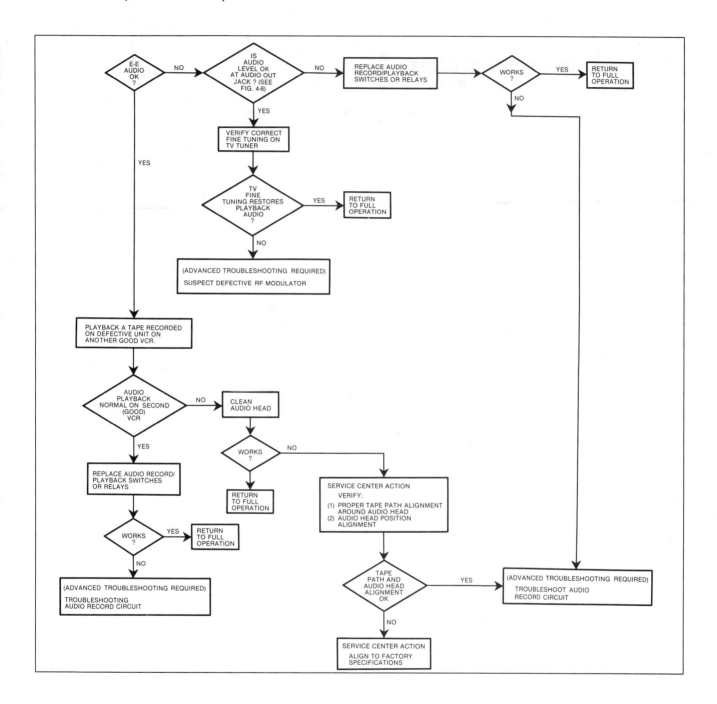

62 VCR TROUBLESHOOTING & REPAIR GUIDE

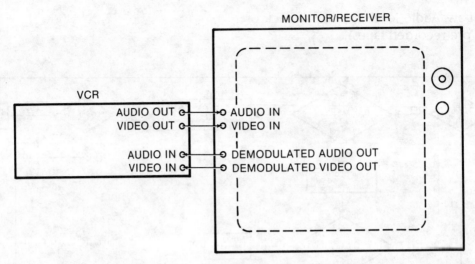

Fig. 4.6. Connect VCR Audio Out to monitor Audio Input and test for proper audio levels from the monitor speaker.

SYMPTOM: Audio plays back at wrong speed.

SYMPTOM: No Audio or Video in playback. Tape moves. E-E Audio/Video O.K.

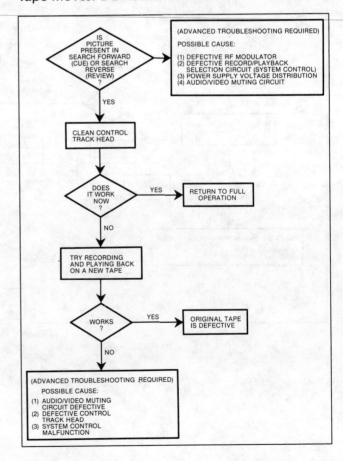

SYMPTOM: Static or popping sound in hi-fi Audio.

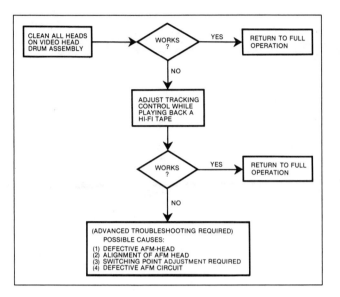

VIDEO PROBLEMS

SYMPTOM: Snow on screen during playback like that shown in Fig. 4.7. Tracking Control won't remove. May or may not have some viewable picture in background. Audio present but may have lots of static.

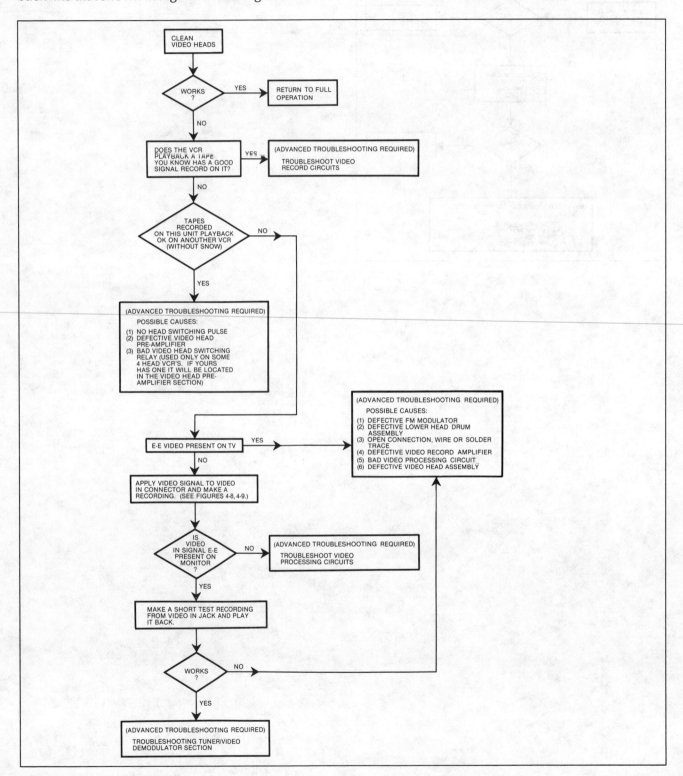

Specific Troubleshooting & Repair 65

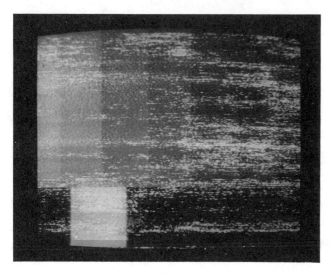

Fig. 4.7. Snow on a display screen.

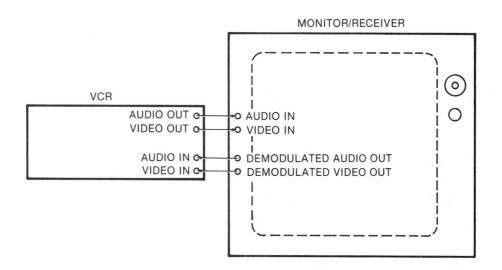

Fig. 4.8. Apply a video signal from the monitor Demodulated Video Out to the VCR Video In and make a recording.

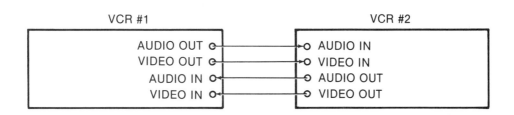

Fig. 4.9. Apply a video signal from VCR 2 Video Out to VCR 1 Video In and make a recording.

66 VCR TROUBLESHOOTING & REPAIR GUIDE

SYMPTOM: Fine horizontal line floats through picture.

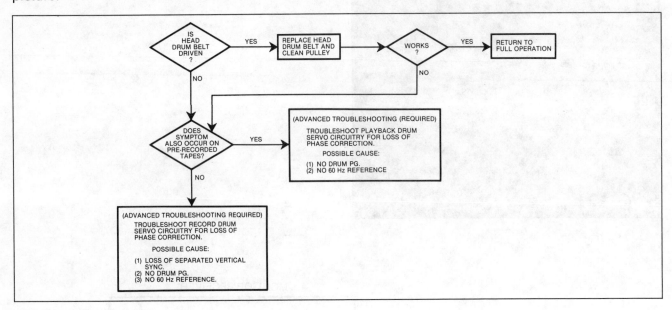

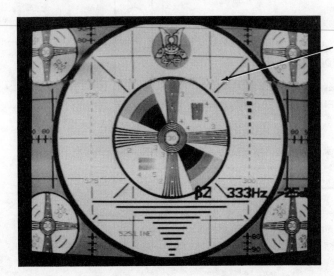

Fig. 4.10. A fine horizontal line can be seen in the top portion of this photograph.

SYMPTOM: Video picture alternates between good video and snow. May occur at regular intervals.

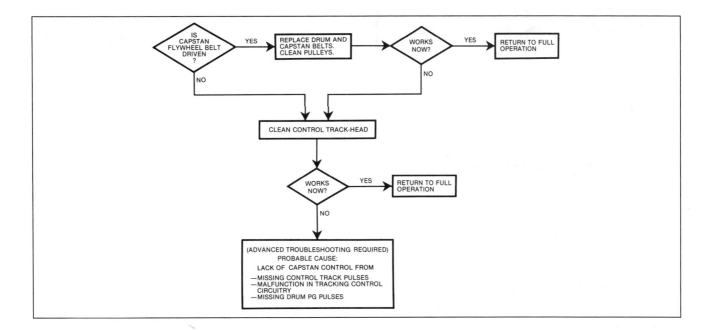

SYMPTOM: No Video or Audio in playback.

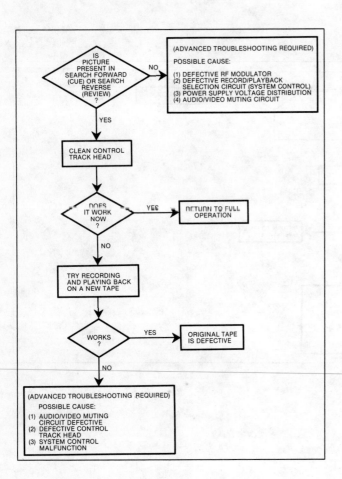

Fig. 4.11. A vertical image bending and tearing at the top of the screen.

SYMPTOM: Vertical images bend or tear near top of screen.

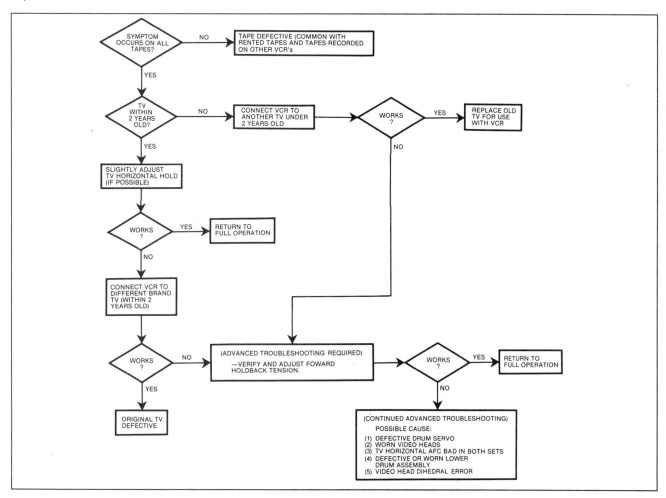

SYMPTOM: Snow bands or lines of interference at top or bottom of screen.

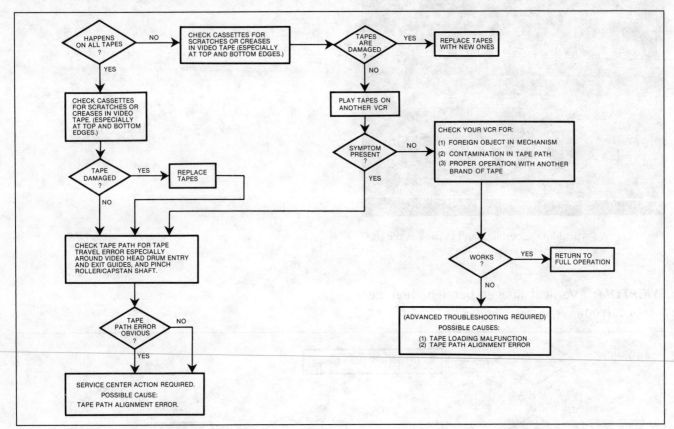

Specific Troubleshooting & Repair 71

Fig. 4.12. Snow bands at the bottom of a screen display.

COLOR PROBLEMS

SYMPTOM: No color on newly recorded tapes. Get color on playback of pre-recorded tapes.

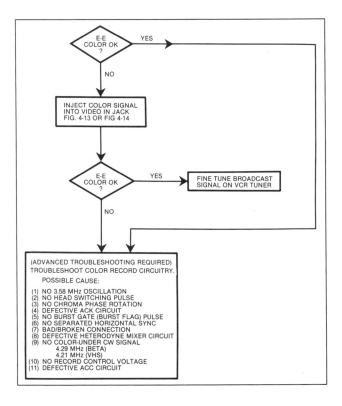

72 VCR TROUBLESHOOTING & REPAIR GUIDE

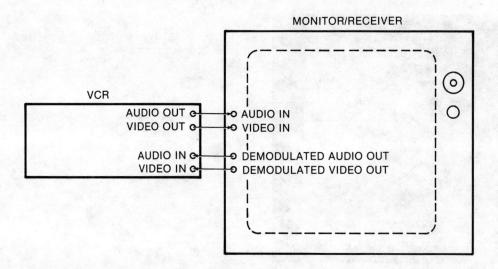

Fig. 4.13. Inject a color signal into the VCR by connecting the monitor Demodulated Video Out to the VCR Video In.

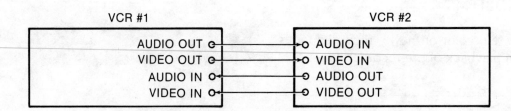

Fig. 4.14. Inject a color signal into VCR 1 by connecting VCR 2 Video Out to VCR 1 Video In.

SYMPTOM: No color in record or playback. Pre-recorded tapes have no color.

SCENARIO

The TV has color when you are viewing normal broadcast signals but no color coming from the VCR when recording or playing back. You are not sure if the VCR is recording color or not. The defect might be in the color playback. In this case the VCR would record color all right but you would never know because it won't play it back.

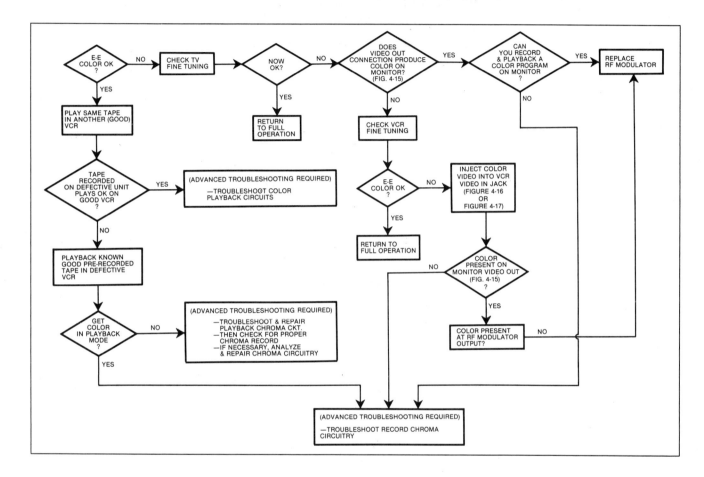

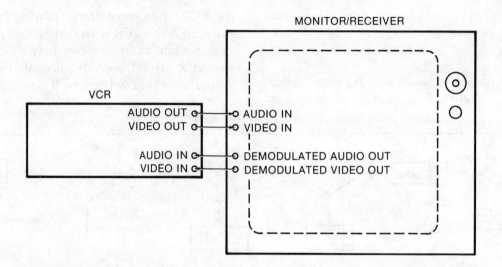

Fig. 4.15. Check for color by connecting VCR Video Out to monitor Video In.

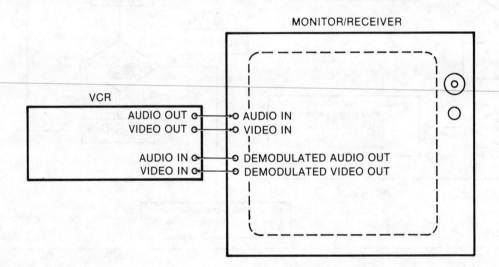

Fig. 4.16. Inject color video into the VCR by connecting the monitor's Demodulated Video Out to the VCR's Video In.

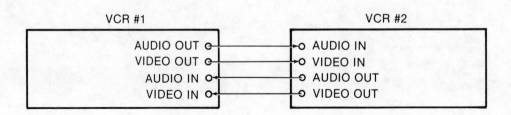

Fig. 4.17. Inject color video into VCR 1 by connecting VCR 2 Video Out to VCR 1 Video In.

SYMPTOM: Bands of color on screen during color playback.

SCENARIO

When you play back a color program on the VCR, bands of color are present on the TV screen. These bands are diagonal and may have the appearance of red, green, blue, purple or a combination of these colors. The bands may show up at unpredictable times, but once they are present they are fairly consistent. This symptom is sometimes known as the "barber pole" effect because of its similarity to the diagonal stripes on a rotating barber's pole. The problem may look like hundreds of bands that are close together on the screen, or there may be only a few spaced farther apart.

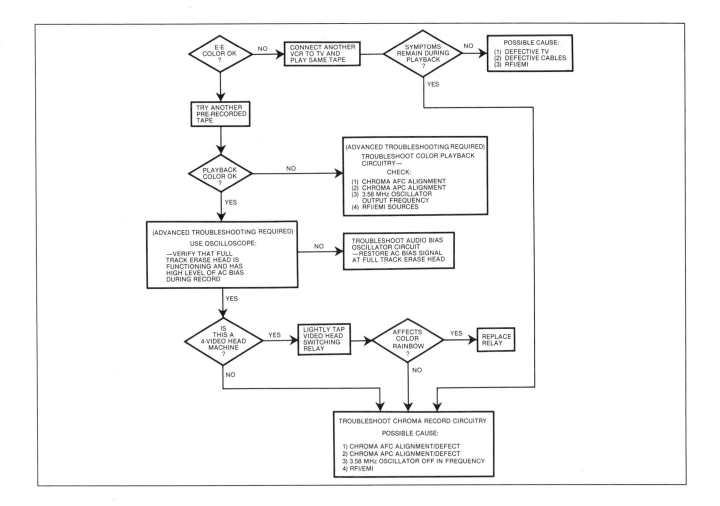

POWER SUPPLY PROBLEMS

SYMPTOM: No functions. Clock may or may not have display.

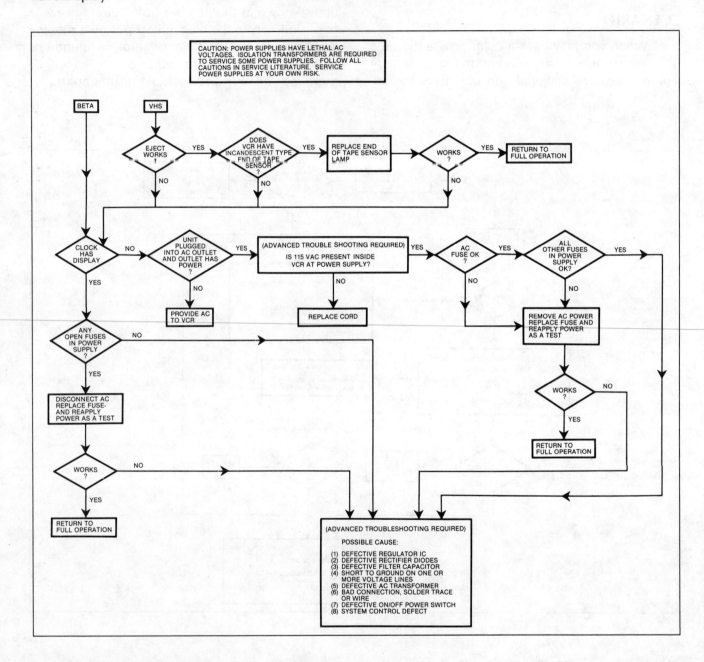

Specific Troubleshooting & Repair

SYMPTOM: No power (Dead VCR).

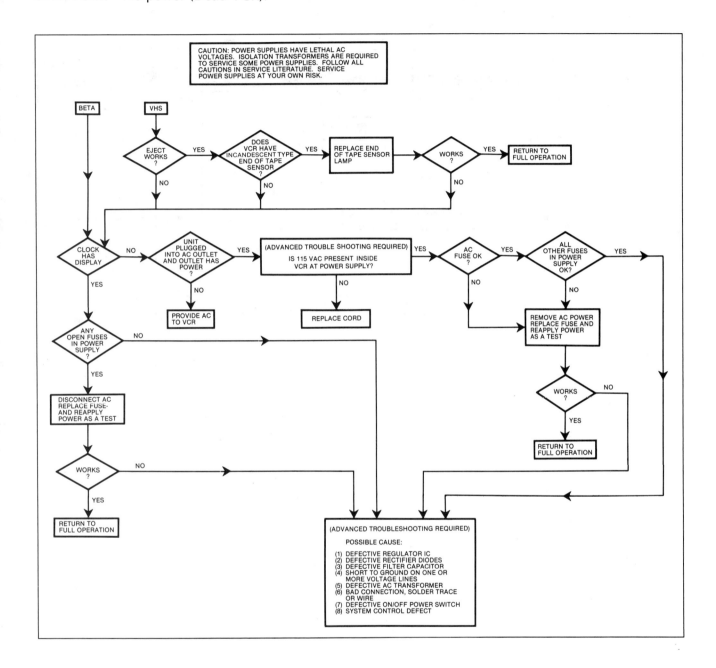

SYMPTOM: No modes functional. Only eject works.

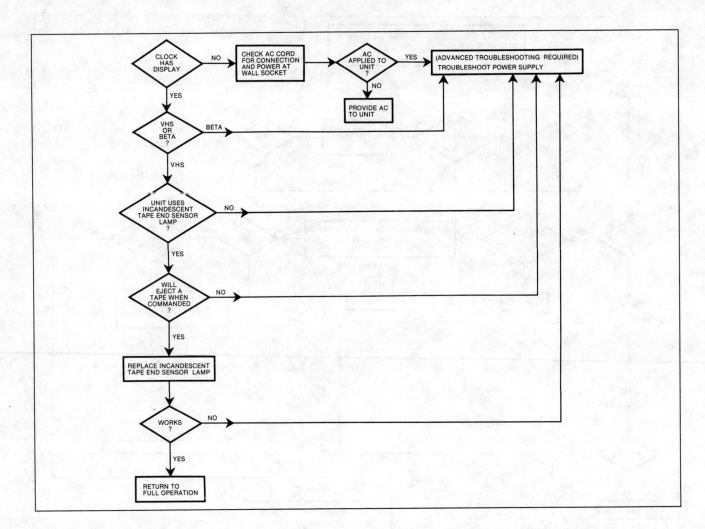

IMPROPER OR NO FUNCTIONS

SYMPTOM: Only eject works.

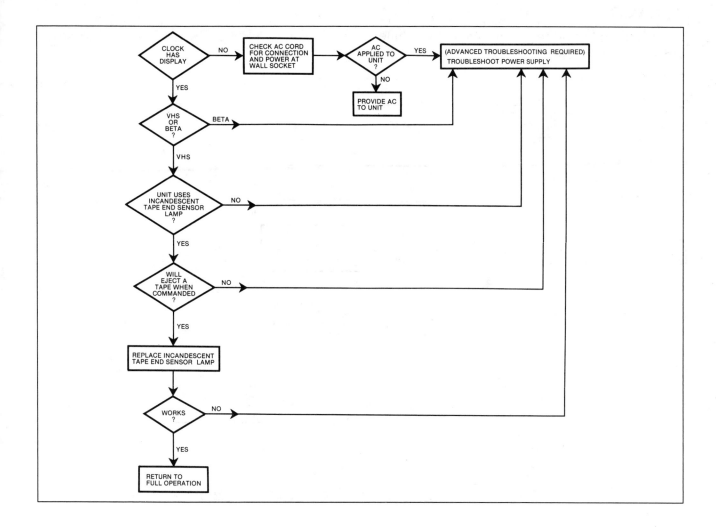

80 VCR TROUBLESHOOTING & REPAIR GUIDE

SYMPTOM: Shuts down or returns to STOP after a few seconds. Tape may spill into unit.

SCENARIO

You have just placed a tape into the machine and pushed PLAY. The VCR makes some of its normal noises for the first few seconds as the tape loads up and then the unit returns to stop. You push the play button again and the same thing happens.

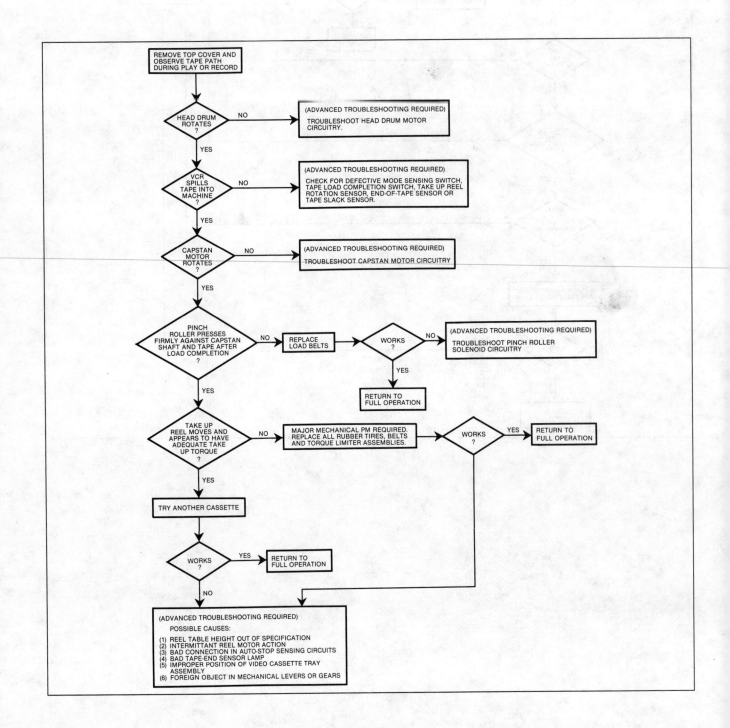

SYMPTOM: Won't properly rewind tape.

SCENARIO

You place a tape in the unit and attempt to rewind it. It seems as though the time required to rewind is a lot longer than when the unit was new, or perhaps the VCR doesn't quite have the power to rewind the tape. Maybe it tries to rewind but stops before the tape is fully rewound. Or it starts rewinding but as it gets to the end of the rewind cycle the tape slows down and stops, even though the motor is still trying to finish rewinding. It is as though the unit is slipping.

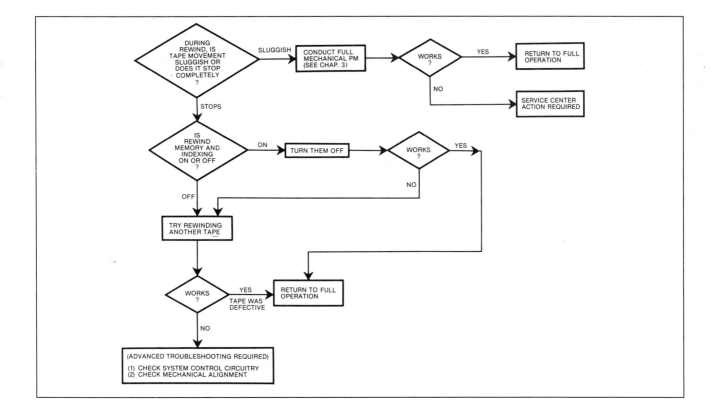

SYMPTOM: Rewind stops before end of tape.

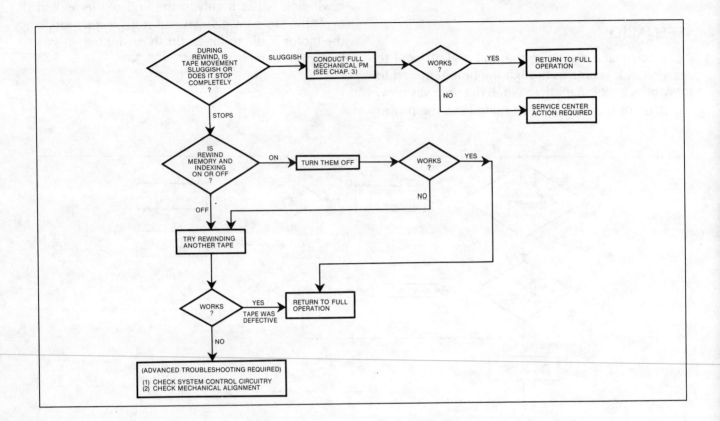

SYMPTOM: Won't rewind at all.

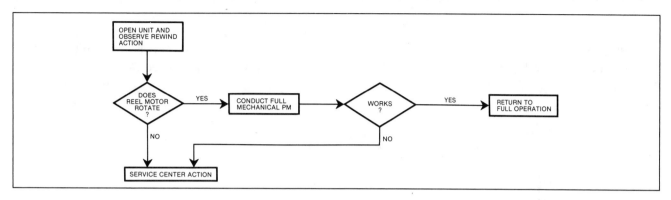

84 VCR TROUBLESHOOTING & REPAIR GUIDE

SYMPTOM: No PLAY, FAST FORWARD, REWIND or RECORD. Here are the possible causes.

1. No power.
2. End-tape sensor lamp out.
3. Mechanical problem (broken belt, slipping tire, etc.).
4. Electrical problem (head drum, capstan doesn't rotate, pinch roller doesn't engage, tape thread timeout, etc.).

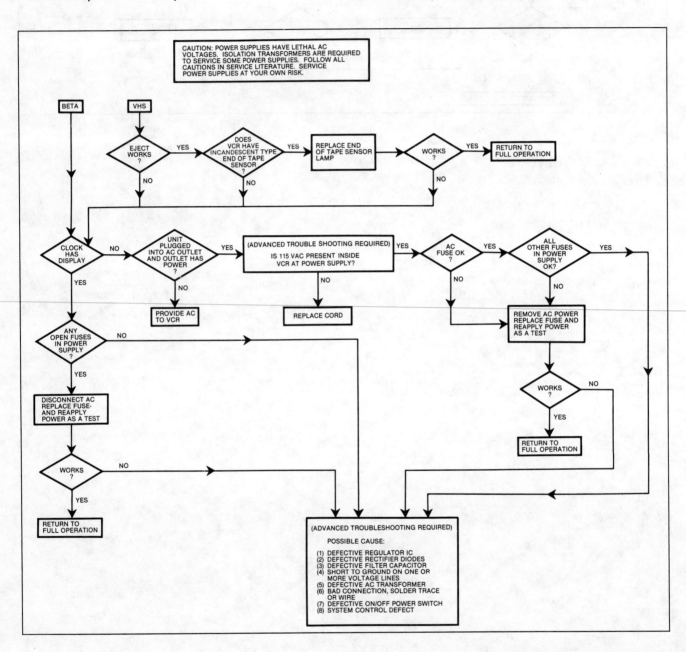

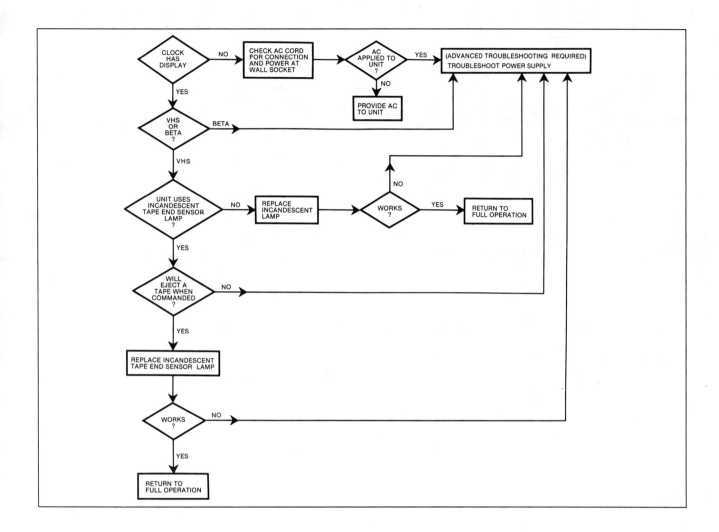

86 VCR TROUBLESHOOTING & REPAIR GUIDE

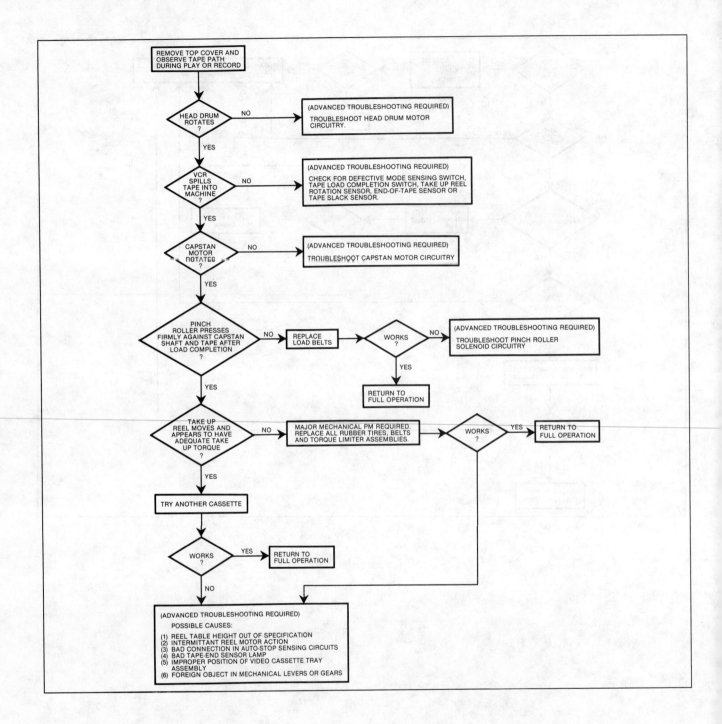

SYMPTOM: Spills tape into the VCR.

SCENARIO

You eject and remove the cassette and discover tape hanging out of the cassette. Some tape could still be caught in the VCR.

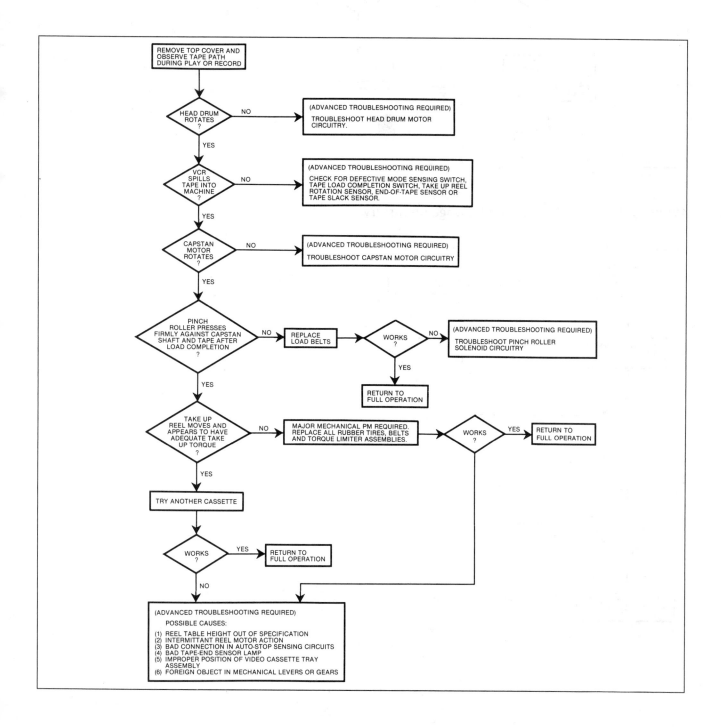

SYMPTOM: Slow or no FAST FORWARD.

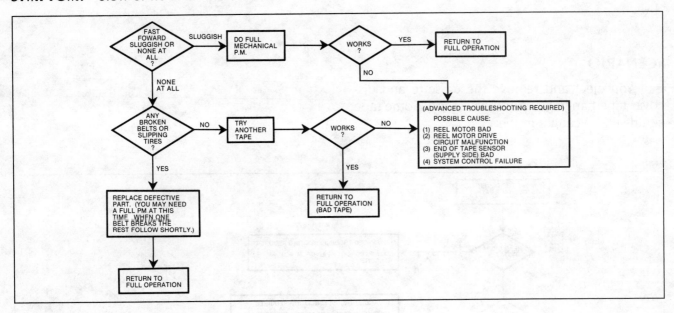

SYMPTOM: Mode button won't engage on first try.

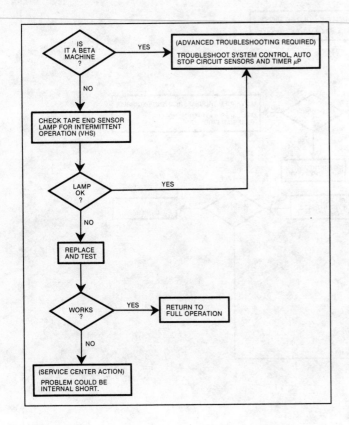

CLOCK OR TIMER PROBLEMS

SYMPTOM: Clock loses 10 minutes each hour.

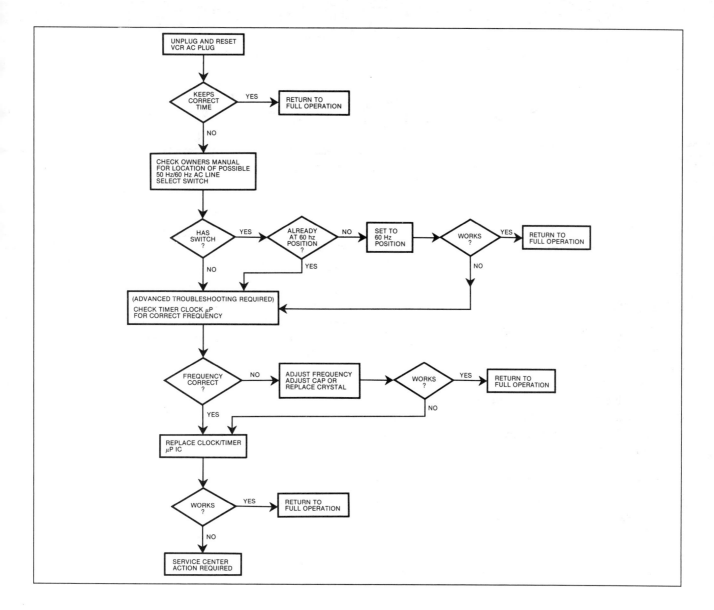

SYMPTOM: No or intermittent record using timer.

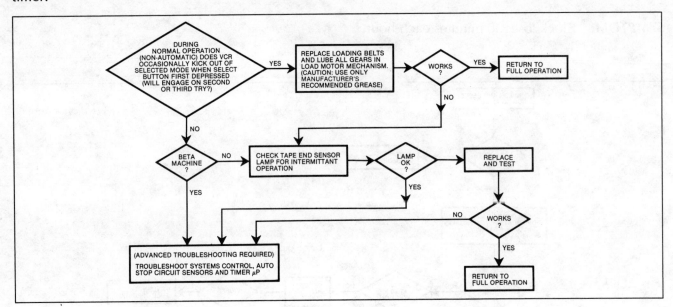

SUMMARY:

This detailed troubleshooting and repair chapter has covered most of the general problems experienced by owners (and servicers) of VCR systems. If following one of the guides in this chapter doesn't solve the problem, you can take the final step (Service Center Action) yourself if you feel qualified. Chapter 7 will provide assistance if you decide to really dig into your machine.

Remember that the information in Chapter 3, "Routine Preventive Maintenance," can help prevent many of the problems analyzed for repair in this chapter.

CAUTION:
Only experienced service technicians should work on power supply and CRT (TV or monitor) displays.

CHAPTER 5

Magnetic Recording Theory

In Chapter 4 you went through a step-by-step guide for detailed troubleshooting and corrective maintenance of Beta and VHS video cassette recorders. Chapter 5 explores the fascinating subject of magnetic recording theory. It presents the technical details of how magnetism is used to record and play back audio and video information.

As you learned in Chapter 1, the technology used to store audio and video signals on a magnetic tape is quite straightforward. Magnetism and the phenomenon of pole attraction have been understood for many years. However, the techniques used to optimize the magnetization of the oxides on a long thin strip of mylar tape vary greatly. The fact that magnetic principles are used remains constant with all forms of recorders.

Most of us have experimented with a simple electromagnet in school science courses. These magnets were made by taking a piece of iron core material and wrapping a copper wire around it. By applying a voltage from a battery or some other power source, this iron core could be made to radiate magnetic lines of force called *flux* which attract and magnetize other metallic objects. When the electrical current flowing through the copper windings increases, so does the strength of the generated magnetic field. VCRs use this same principle to record video information on a magnetic tape. In a VCR, a head becomes the iron core and a thin film of ferris oxide pasted on a flexible mylar or plastic tape becomes the object to be magnetized.

VIDEO HEAD CONSTRUCTION

The video head is a very tiny electromagnet made of a material that can be easily magnetized. Such a material is called *permeable* and is symbolized by the Greek letter μ. Permeable material has magnetic domains that align with the flux of an external magnetic field. Shaped loosely like a horseshoe with a tiny gap at one end, the head is mounted so it barely makes contact with a flexible tape at the point of the gap as shown in Fig. 5.1.

The size of a video head is compared with a one cent coin in Fig. 5.2.

Magnetic fields or flux lines radiate from this gap into a nearby tape surface. Flux extends across the gap between the two poles on the video head core. Thus, this core material adds its own magnetic influence to the external signal and increases

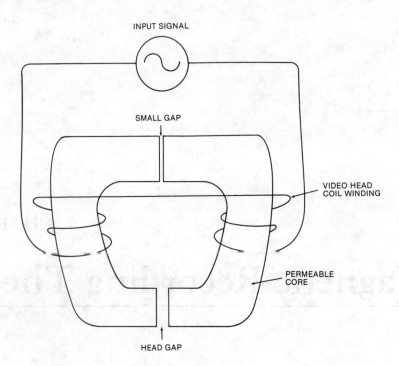

Fig. 5.1. Video head construction.

Fig. 5.2. Photograph comparing a video head and a coin.

the intensity of the field. The shape of the head and the core material itself affect the permeability of the head. See Fig. 5.3.

With electrical current flowing through the wire coil, the head gap is moved with respect to the video tape. This movement causes magnetic particles on the surface of the tape to align so that the tape coating becomes magnetized in an arrangement determined by the current of the electrical coil.

The tape, constructed of a polyester base, is thinly coated with a binder material that contains very small metallic particles, typically iron oxide. The material tends to retain the alignment of the particles after magnetic action. This ability to retain magnetic flux after an external magnetizing force has been removed is called *retention*. Thus, the tiny metallic particles on the video tape retain their magnetic alignment (*polarities*) long after the tape has passed the recording video head.

The flux of the video head is concentrated and focused at the gap by the design of the head itself to produce a maximum field intensity at the point where the head gap contacts the tape. If an alternating signal is applied across the coil of the video head, the flux field produced will magnetize the tape at the same alternating rate producing columns of particles aligned north to south, north to south, etc. The recording current passing through the coil of wire wrapped around the video head material produces a magnetic flux field that alternates proportionally to the signal. This alternating current is essential for recording intelligent information on the tape.

The tape then is moved relative to the head as an electric current in the coil modifies the magnetic field in the core and head gap area. If the tape and head were to remain stationary with respect to each other, the magnetic polarization

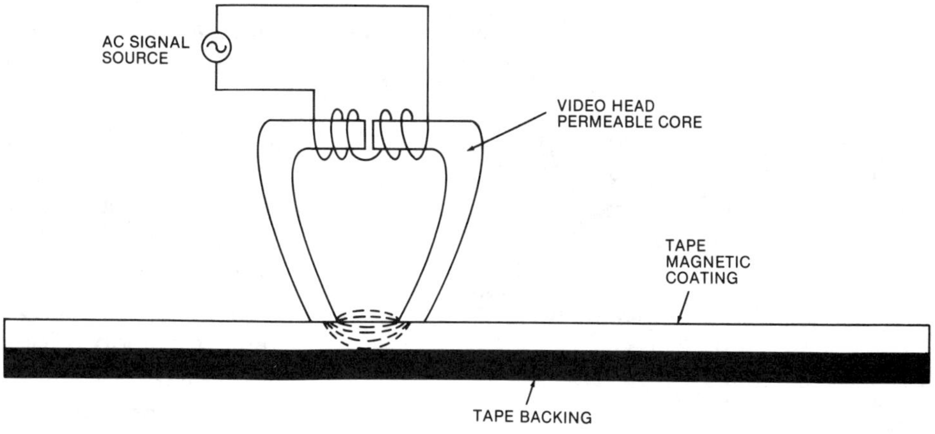

Fig. 5.3. Magnetic lines of flux radiate from the high reluctance head gap through the magnetic coating of the tape.

would vary in only one location on the tape and a single spot would be constantly magnetized in alternating directions (polarities).

Careful consideration is given in selecting the material of the video head permeable core to minimize resistance to magnetic flux. The resistance of a material to align magnetically with the influence of an external field is called *reluctance*. An analogy is the resistance of some materials to the flow of electrons in electronic theory.

Other losses result from circular flows of electrons called *eddy currents* which occur in the core material itself as it is exposed to rapidly changing magnetic fields. Eddy currents behave like short circuits and dissipate heat energy in the core. Eddy currents increase with the increase in signal frequency applied to the head coil. As these currents increase, they resist the magnetic field being applied to the oxide on the tape producing heat in the core material and weakening the strength of the flux field focused at the head gap.

To minimize these losses, the video head core is constructed of very poor conductive material, typically ferric oxide mixed with oxides of other metals such as zinc, magnesium, and nickel. Collectively called *ferrite*, this crystalline material is characterized by a high resistance to electrical conduction. It resists current flow. Ferrite heads are more efficient at high frequencies so their characteristics are ideal for the frequencies associated with video tape recording.

The core is typically constructed from two pieces of material. At one end, the core pieces are tightly connected to minimize reluctance. The other end of the core has a gap filled with a non-magnetic material which maintains the critical width of the gap. It's also important to control the thickness of the core material itself (in the area of the gap). The head gap material on many video heads is silicon dioxide (glass).

Because the video head core is highly permeable with a gap characterized by high reluctance, the magnetic lines of flux actually flow into and through the tape surface, which is also highly permeable.

MAGNETIC TAPE RECORDING PRINCIPLES

The process of storing and retrieving magnetic video tape information is not as easy as it may first appear. Several phenomena have major effects on the magnetic recording process. These include the composition and thickness of the metallic coating on the tape, and the strength and frequency of the magnetic signal applied through the recording head.

Limitations of Video Recording

Some of the limitations of video recording are shown in the BH curve in Fig. 5.4. The BH curve is a graph showing magnetic tape characteristics during recording and the ability of the tape to retain the information once the external magnetic influence has been removed. The remaining flux density on recorded tape is designated "B" and is measured in nanowebers per square meter.

The horizontal line represents magnetic field intensity, H. This intensity is measured in ampere turns per meter. It's the amount of current (in

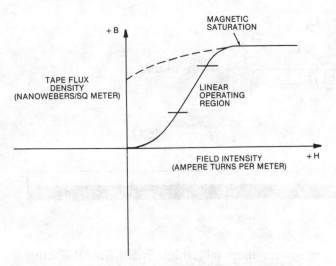

Fig. 5.4. Magnetic characteristics of recorded tape.

amps) applied directly to the coil windings on the video head. As shown on the graph, flux density, B, increases in direct proportion to the field intensity caused by an increase in video head coil current. At very low record current levels, little flux density remains after the coil current is reduced to zero. This is because residual magnetism attempts to keep the atomic structures polarized in their previous orientation. This is a form of reluctance. The atomic structures resist any force to realign the particles in the same direction as the externally applied field. This low record current is enough to maintain some residual magnetic influence on the tape.

As the coil current increases, the flux density begins to increase linearly. A point is reached where all the magnetic oxide particles on the tape are oriented to the same polarity as the externally applied field and current increase no longer produces a change in flux density. This is called the *magnetic saturation point* and is represented by a flat horizontal line on the graph.

Fig. 5.4 also shows the effect of record current on field intensity once the recording signal level has been reduced to zero. As indicated by the dotted line on the graph, as the external field is reduced, the flux density level does not return to zero but drops to a level slightly less than it was when the external field was being applied through the video heads.

To obtain zero flux density, the current passing through the coil of a video head core must generate a magnetic field in the head gap area of opposite polarity. The field intensity necessary to produce zero flux is called *coersive force*. As coersive force increases, magnetic polarization is driven negative until it eventually reaches the magnetic saturation point of opposite polarity as shown in Fig. 5.5. If the current in the video head coil is made to alternate at equal amplitudes in both the positive and negative directions, a closed hysteresis loop is formed as shown on the graph. The magnetic coating on the video tape must have high magnetic retention and should reach high flux densities before saturation. These characteristics are important in evaluating the differences between normal and high grade video tapes.

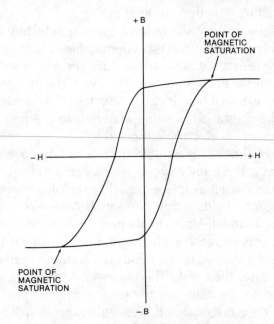

Fig. 5.5. Hysteresis loop formed by alternating currents of equal amplitude.

Audio tape recorders are designed so they operate in the linear region of the BH curve. In this area, the relationship between flux density and the applied signal current is uniform. If an audio signal is applied directly to the head, and the strength of that signal varies between zero amps and the amount required for saturation, the playback signal read off the tape will be distorted. If, however, the incoming signal varies in the linear region on the BH curve, the playback signal closely represents what was recorded.

Conventional audio recorders use a high

frequency AC signal known as *bias* to cause linear BH operation. A bias signal is added to the audio signal in the head just before it is recorded on the tape to insure that the audio signal level is well within the linear region on the BH curve shown in Fig. 5.6.

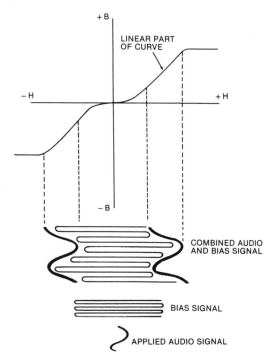

Fig. 5.6. Record bias causes signal recording only in the linear region of the BH curve.

Bias frequencies range from 60 kHz to almost 100 kHz. The VCR, however, doesn't use bias in recording video information in the same way as an audio tape recorder. Linear operation is accomplished using a method that will be discussed later.

The Importance of Wavelength

Having described one of the challenges of recording information onto a tape, it's appropriate to understand wavelength and the term *recorded wavelength*. Wavelength is the inverse of frequency and is measured in meters. It represents the distance that one cycle of a signal travels in one second. Recorded wavelength is the length of tape required to record one full cycle of alternating current signal.

Wavelength, lambda (λ), is determined by the speed of the tape and the frequency of the signal being recorded. For example, if a frequency of one hertz (cycles per second) is recorded on a tape traveling at a speed of ten inches per second, the recorded wavelength would be ten inches as shown in Fig. 5.7.

As this one hertz signal is recorded, the polarity of the magnetic field across the head gap is initially polarized in neither the north or south direction. As the tape passes across the head and the applied signal becomes more positive, the poles on the head become oriented toward the north. After two-and-a-half inches of tape travel beneath the head, the applied AC signal has peaked and begins to decrease. The magnetic lines of flux across the head also decrease. At the five-inch mark on the tape, the AC signal reaches the zero flux point and starts in the negative direction. The head gap polarities begin to reverse and the recorded field starts shifting to the south. The applied AC current finally reaches the zero point again at the ten-inch mark. Playback of the recorded signal is accomplished by rewinding the tape and then pulling it across the heads again which causes the magnetic field on the tape to generate a tiny alternating current in the head.

As described in Fig. 5.8, maximum playback output occurs when the head gap width is half that of the recorded wavelength. Once the recorded wavelength signal reaches the one-half point, the output from the playback head suddenly drops to zero within the next octave. This is easy to understand from the following illustration. If both the positive and negative portions of the recorded wavelength signal are present in equal amounts within the head gap boundaries, the north and south poles cancel each other and the resultant output signal from the video head is zero.

In playback, the ability of the video head to detect magnetic lines of flux on the tape is not equal over all frequencies. During playback, the recording circuit is turned off and a high-gain amplifier is connected to the windings of the video head coils through a rotary transformer. As the magnetized tape is pulled across the head gap, the magnetic flux on the tape seeks the path of least reluctance. This path is through the core of the video head. As the magnetic fields expand and contract in the core, a voltage is produced in the windings on the video head. The magnetic field on the tape must be

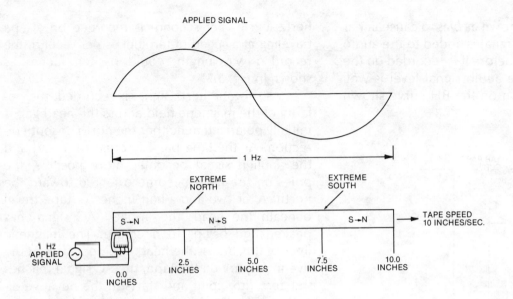

Fig. 5.7. Recorded wavelength (calculated using frequency and tape speed).

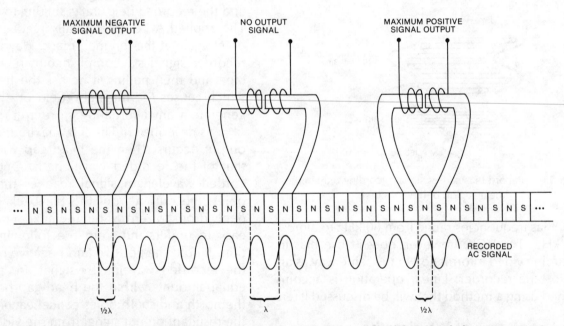

Fig. 5.8. Maximum signal occurs when the head gap equals half the recorded wavelength.

changing to cause an output from the video head coils as shown in Fig. 5.9. Voltage is only produced in video head coils when magnetic lines of force intersect the coil windings and this only occurs when a magnetic field is expanding or contracting.

To observe the effect that the change in flux has on a head during playback, observe the record and playback control track signals within the VCR itself. The purpose and theory of control track signals will be discussed in detail later, but it's useful to introduce them here to illustrate an important point.

During the recording process, a 30 Hz square wave signal is fed into a stationary control track head that writes the signal on the tape. When this signal is retrieved off the tape, the 30 Hz square wave signal is distorted. The playback signal coming off the control head appears to be a differentiated square wave with a series of positive and negative going spikes as shown in Fig. 5.9. This

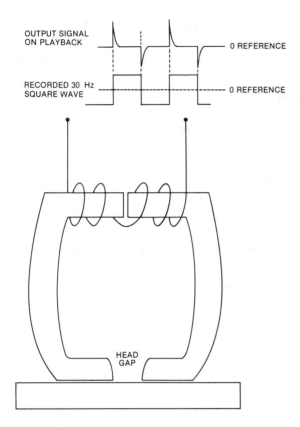

Fig. 5.9. Output playback signal occurs only when magnetic lines of flux change at the head gap.

waveform occurs because the square wave signal is recorded directly on the tape via the control track head using no AC bias.

During playback, the positive half of the recorded magnetic flux is aligned uniformly and points in one direction. During the next half cycle, the recorded flux is equal in strength but points in the opposite (negative) direction.

Playback distortion is also caused by the changing flux in the core when the head gap encounters a place on the tape where the flux has also changed. This core flux change occurs only at the positive going or negative going excursions of the square wave so head output only occurs where the lines of flux have changed.

The strength of the signal coming off the video head is directly proportional to the frequency at which the magnetic flux changes. If the playback frequency is doubled, the flux rate is doubled. The resultant output signal from the video head is also doubled. The doubling of the playback frequency is described as a *one octave*

increase. The doubling of the output voltage is a rise of six *decibels* (dB). A decibel is a measurement of signal strength.

Typical video head playback characteristics are six dB per octave, so voltage doubles every 6 dB as shown in Fig. 5.10. The video head output signal increases at 6 dB per octave until the rate of change begins to decrease. At that point the signal voltage quickly begins to decrease, reaching zero at twice the frequency of maximum output. The frequency response levels off and then quickly decreases because hysteresis and eddy current losses have increased effect on recorded wavelength at higher frequencies. This effect is caused partly by thermal losses from shock waves at the magnetic domain boundaries. These shock waves result from changes in atomic dimensions caused by electron spin. Head gap also limits high frequency response.

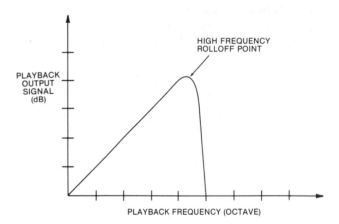

Fig. 5.10. Playback signal increases 6 dB per octave.

Equalization

Conventional audio recorders can overcome the nonlinear playback frequency response by the use of *electronic equalization* and by using head gap widths and record wavelengths that are well under the high frequency roll-off point of the head.

Electronic equalization decreases the playback head amplifier gain as the playback signal frequency increases. This process compensates for the increased output signal from the playback head by reducing the electronic amplification at high frequencies. The response curve for the equalization circuit resembles a mirror image of the playback frequency response characteristic curve. Low frequencies are amplified or boosted while high frequencies are reduced. The resultant output is a flat frequency response.

Equalization is not a difficult problem for design engineers. However, there are limitations which restrict its success. One of these is associated with the dynamic range of the tape itself. *Dynamic range* is the ratio of (a) the largest signal that can be recorded on the tape before it reaches magnetic saturation to (b) the lowest level signal that can be detected above the noise threshold. Tape noise is also called *hiss*.

Tapes available today have a dynamic range of approximately 70 dB. This permits slightly more than 10 octaves of bandwidth for recording signals directly onto the tape. Good quality audio recorders can process signals from about 20 Hz to 20,000 Hz (20 kHz). The frequency response of these recorders spans approximately 10 octaves using a direct recording process with audio record bias.

The frequency range for video signals is from about 30 hertz to above four million hertz (4 MHz). This span of more than 16 octaves is well above the dynamic range of magnetic tape. Equalizing such a signal is essentially not possible.

However, recording the wide range of video frequencies can be accomplished by reducing the head gap width, increasing the speed of recording (writing speed), and recording the signal with a process called *frequency modulation*.

HEAD GAP

The head gap on a stationary head of a conventional audio recorder is typically several microns (millionths of a meter) wide. Since video tape recorders deal with much higher frequencies, the gap on a video head is measured in tenths of microns. Video head gaps in consumer ½-inch VCRs range between 0.3 and 0.6 microns. Reducing the thickness of this gap results in a loss of sensitivity because the reluctance of the gap drops as the two parts of the core (poles) approach each other. There are other factors which cause losses in video record and playback heads. The resistance of the coil windings, frequency-dependent capacitance effects, and head construction also cause signal level losses in video heads.

A video head coil with only a few turns will have low resistive and capacitive losses, but it will also have lower playback output voltage. If many turns of wire are used, the playback voltage is increased, but the coil can have so much stray capacitance that the head's own inductance will cause the head to resonate (vibrate). It's possible for the resonant frequency to be lower than the highest frequency of the signal to be received and played back. If this occurs, severe signal loss results at higher frequencies. This situation requires a compromise in the production and design of the video head. Typical video head coils have from 1000 to 2000 turns of wire with a coil wrapped around both halves of the head core.

Writing Speed

Conventional audio recorders utilize stationary heads to record and play back audio program signals. This can be accomplished using relatively low tape speed and a relatively wide head gap because audio frequencies are limited to a small (20 Hz - 20 kHz) frequency range.

Video signals require complex design solutions. The luminance or brightness portion of a video picture ranges (in theory) from zero to about 5 MHz. Some of the first units (e.g., RCA's 1954 product line) were designed using stationary video heads.

Reproducing a signal with a recorded wavelength that covers the video frequency spectrum, using the stationary video head system, requires a tape speed of approximately 30 feet per second. As shown by the equation below, to produce an hour-long video program on this machine you would need over 20 miles of tape.

$$\text{LENGTH} = \frac{30 \text{ feet}}{\text{sec.}} \times \frac{60 \text{ sec.}}{1 \text{ min.}} \times \frac{60 \text{ min.}}{1 \text{ hour}} \times \frac{1 \text{ mile}}{5280 \text{ feet}}$$

$$\text{LENGTH} = 20.45 \text{ miles/hour}$$

The tape must pass the stationary video head at a speed of twenty-and-one-half miles per hour to play a single hour of program.

The approximate tape speed required to write (record) a video signal on a tape can be calculated by multiplying the frequency of the record signal (in hertz) by the wavelength (in meters) as shown b

[text obscured by library receipt]

$$L = \frac{5.0 \text{ meters}}{\text{sec.}} \times \frac{\text{...}}{1 \text{ min.}} \times \frac{\text{...}}{1 \text{ hour}} \times \frac{3.14 \text{ feet}}{1 \text{ meter}}$$
$$\times \frac{1 \text{ mile}}{5280 \text{ ft.}}$$

$$L = 123.1445 \text{ miles}$$

This is obviously unacceptable.

One way to reduce the amount of tape required to record a video program is to reduce the width of the video head gap. However, there are practical limits to the size of the video head gap. An alternate solution is to design a high speed relationship between the video heads and the tape while reducing the actual speed of the tape traveling through the machine. This head-to-tape relationship is called *writing speed*. It's achieved by moving the tape through the machine at a relatively slow speed (slower than an audio cassette player) while rotating the drum assembly containing the video heads at high speed (about 1800 RPM).

The electrical energy for record and playback is transmitted to and from the video heads via a transformer that rotates with the drum. The tape is drawn around the drum assembly at a slight angle using precisely positioned tape guides. The tape wraps slightly more than 180 degrees around the drum in what is called *helical-scan recording*. As the video head assembly (also known as a *scanner*, *upper cylinder*, or *upper drum*) rotates, the heads on the drum contact the relatively slow moving video tape and produce an equivalent writing speed of around 250 inches per second (ips). The end result is a stretched wavelength enabling high frequency recording without high frequency roll off.

A head gap size is used that is large enough to provide adequate sensitivity. In two-head helical-scan recording, two heads are used to record one full frame of video. They are mounted on a rotating head drum assembly. Fig. 5.11 shows a Beta head drum with the stationary upper drum removed. Each head records one-half of the frame.

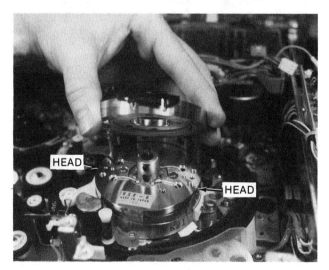

Fig. 5.11. Two heads are mounted on a rotating head drum assembly.

If you compare a video frame to a single frame of movie film, you will notice that if the movie film is stopped at a particular frame, an image is projected on the screen that resembles a 35mm slide picture. Movie film is comprised of a series of still pictures which are moved past a projection lens at a rate of 24 frames per second. The velocity of this motion is just fast enough to cause our brain to sense continual motion in the projected image without flickering or discontinuity. This is analogous to the image from a video recorder system.

The video picture on a TV screen is divided into 30 frames. Each of these frames is further divided into 525 horizontal lines. Line 1 is at the very top of the TV screen and line 525 is at the very bottom of the screen. The 525 lines are categorized into two fields. Even-numbered lines make up one field; the odd-numbered lines make up the other field. To reduce flicker on the TV screen, these fields are projected onto the face of the TV cathode ray tube (CRT) screen at alternating intervals. The field containing the odd-numbered lines begins by scanning the odd lines on the CRT from left to right and top to bottom. As soon as this is completed, the beam is quickly returned to the top of the screen and the even-lines are scanned onto the screen. This procedure is known as *2 to 1 interlace*. The horizontal lines of video are interlaced on your screen using a procedure that is repeated 59.94 times per second (vertical field rate). Since each field contains 262.5 horizontal lines, one frame contains two vertical fields or 525 horizontal lines.

The two-head helical-scan video recorder uses two heads mounted exactly 180 degrees apart on the drum. Fig. 5.12 shows one of two video heads mounted on a Beta drum. Fig. 5.13 shows the video heads mounted on a VHS head drum.

One head records the first field and the other head records the second field. With the video tape wrapped slightly more than 180 degrees around the head drum, an extra three horizontal lines of video are recorded from the alternate field at the beginning and end of each sweep of the head. This produces six lines of identical video information during playback. The two video heads have the record signal simultaneously applied at all times.

Calculation of writing speed in consumer helical-scan recording recognizes that the heads are passing over the tape in the same direction that the tape is traveling and considers the diameter of the scanner. The fact that the heads strike the tape at a slight angle (about five degrees) has little effect on

Fig. 5.12. Beta video head drum.

the speed. Other considerations include manufacturing tolerances, the amount of lower FM sideband reclaimed and miscellaneous high frequency loss.

If the tape didn't move, the length of the video head track would be half the circumference of the head drum. (For simplification, assume the tape is wrapped exactly 180 degrees around the scanner). The scanner (video head assembly) diameter in a Betamax system is 74.487 mm. Using the equation for circumference, C:

$$C = 2(\pi) \times (r)$$

where
$\pi = 3.1416$,
r = radius (½ diameter),
C is calculated: 74.487 mm $\times$ (3.1416)/2 = 117.04 mm

For simplicity, assume that the five-degree track angle has an insignificant effect on the equation so can be ignored and that the video head track is parallel to the tape run. Since the heads

Fig. 5.13. VHS video head drum.

record one field in half the circumference, multiply one-half the circumference by the vertical field rate and subtract the linear speed of the tape because the tape travels in the same direction as the head rotates.

Using the above values, Writing Speed is equal to the diameter times π divided by two times the vertical field rate less the linear tape speed.

Writing Speed = 117.04 mm × 59.94 − 40 mm/sec. = 6975 mm/s or 6.975 m/sec.

The published specification for Betamax Beta I writing speed is 6.973 meters per second.

The writing speeds for Beta II and Beta III recording are slightly slower because only the linear speed of the tape changes. The linear tape speed for the Beta II is 20 mm/s (0.8 ips) and 13.3 mm/s (0.52 ips) for Beta III.

VHS machines have a smaller head drum diameter (6.2 centimeters) and slower linear tape speeds of 33.35 mm/sec. (1.3 ips) for SP, 16.7 mm/sec. (0.66 ips) for LP and 11.12 mm/sec. (0.4 ips) for SLP or EP. Writing speed for the VHS machines in SP is 5.8 m/sec. (228 ips).

One reason the Beta format produces a sharper picture is its higher writing speed. The higher the writing speed, the more detail that can be reproduced in the record/playback process. VHS manufacturers chose a lower writing speed to conserve tape. Thus while longer recording times are a selling feature of VHS units, these machines sacrifice reproduction quality.

FM RECORD/PLAYBACK

The final difficulty to be overcome in achieving suitable video recording is the dynamic range limitation of the tape itself. The typical dynamic range of currently available tape is slightly more than 10 octaves. To record a signal ranging from 30 Hz to 4 MHz requires over 16 octaves.

The tape dynamic range problem was overcome by using frequency modulation (FM) for recording the video luminance signal. Frequency modulation means that a certain frequency corresponds directly to the amplitude of the original information coming in from the TV transmitter. Video information broadcast from a TV station is measured in units of amplitude and consists of many complex signals. These signals include color or chroma information, brightness or luminance information (Y), horizontal and vertical synchronization pulses (*sync*) and a short, high-frequency oscillation signal (*burst*) that is used to synchronize the color circuits of the receiving device with those at the TV transmitter.

For the moment, consider the luminance or Y signal as containing the brightness information, picture detail and the horizontal and vertical synchronization pulses. Horizontal sync pulses phase or frequency lock a receiver's horizontal circuits with those from the broadcast station.

In a TV receiver, these horizontal synchronization pulses are responsible for horizontally locking the picture to the broadcast signal. Vertical sync pulses perform a similar role in the vertical circuitry of the TV and prevent the TV picture from rolling in a vertical direction. Vertical sync is also used in the servo circuitry of the VCR.

The Y portion of the TV signal is frequency

modulated with a 3.5 MHz to 4.8 MHz signal for conventional Betamax and 3.4 MHz to 4.4 MHz signal for VHS. The FM frequency deviation varies depending on whether or not the unit uses VHS, Beta or SuperBeta format.

In conventional Beta format, the horizontal and vertical sync tip pulses correspond to the 3.5 MHz portion of the frequency spectrum. Maximum brightness in the TV image corresponds to 4.8 MHz.

As the picture brightness varies from one part of the TV screen to another, the signal frequency changes from one frequency to another. Frequency is directly related to the amplitude of the original picture signal. Frequency modulation produces many complex sidebands which extend above and below the deviation frequency. These sidebands are also recorded on the tape. Most of the signal energy in the sidebands is in the frequencies closest to the main signal. During playback, enough sideband information is recovered to produce a close replica of the original video signal. In Betamax machines, for example, the sideband information extends from approximately 1 MHz to 5.1 MHz as depicted in Fig. 5.14. A bandwidth less than three octaves is required to reproduce this signal.

The signal can be easily equalized in this range, overcoming the difficult task of equalizing over 16 octaves of the original video signal.

The use of FM overcomes another obstacle associated with video recording. A video signal can contain DC information. For example, if a blank picture (having no detail) and covering the entire screen (such as a blank white piece of paper) were transmitted from the TV station, the video signal would contain no alternating current information. This represents a constant DC signal. If this DC signal were recorded on tape using conventional recording methods, no playback information could be retrieved from the tape. For video heads to extract tape information, the magnetic lines of flux must be changing polarity. However, if this DC signal is converted to a frequency that represents that DC signal, it can be recorded on tape, played back, and converted once again to the original DC signal. This is accomplished using FM recording.

FM recording is ideal because it solves tape dynamic range limitations, and the system can be designed to ignore minor playback amplitude variations of the FM signal caused by imperfections in tape magnetic coatings and mechanical errors in the machine. During playback, the VCR monitors variations (deviation) from a set modulation frequency. Upon detection, the machine converts these signals into the original received signal. Because the system detects variations in frequency and not amplitude, it isn't sensitive to minor ampli-

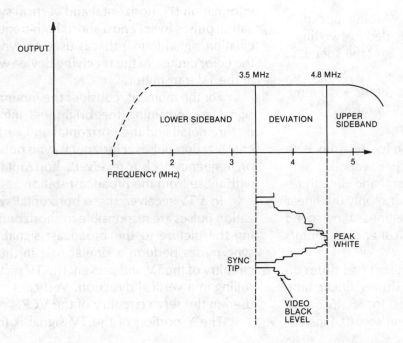

Fig. 5.14. Frequency spectrum for FM Beta video includes a narrow 3.5–4.8 MHz deviation band and upper and lower sidebands.

tude variations in playback signal strength. The FM process also permits the recorded signal to saturate the video tape without causing adverse playback video signal distortion. This permits the FM signal to be directly recorded onto the tape without using bias. During playback, the FM signal coming off the tape is distorted, but, because the FM demodulator is only looking for variations in frequency, the original signal is restored. The ability to saturate the tape during recording produces stronger playback signals, and decreases the chance that tape or VCR mechanical imperfections could affect the playback signal.

As mentioned, Beta and VHS units using two-head helical-scan have the video tape wrapped around the head drum assembly slightly more than 180 degrees although the video heads are mounted exactly 180 degrees apart. The tape is wrapped just far enough around the head drum assembly that one head is starting to exit the tape after the other head has made contact as shown in Fig. 5.15. The record FM signal is constantly applied to both heads simultaneously so there is a duplication of recorded material on the tape for slightly more than three horizontal lines. As described in Fig. 5.16 when video head B is starting to exit the tape, it is recording the same information as video head A which is just starting to touch the tape on the other side of the drum. This duplication of recorded material exists for three to four horizontal lines, yet the recording process causes the video head passing over the tape first to make contact approximately 10 horizontal lines before the vertical sync period occurs.

During playback, the video heads are electronically switched on and off to produce a normal picture and are timed to avoid playing back the duplicated three horizontal lines as shown in Fig. 5.17. The video heads must be switched on and off during playback to prevent video noise generated by the head which is not in contact with the tape. Each video head is connected to its own high-gain amplifier called a *pre-amplifier*. If the pre-amplifier isn't de-energized when the video head is not in contact with the tape, it would produce electronic noise that would appear as snow on the screen. Only the amplifier for the head in contact with the tape should be active. Video head switching occurs 6½ horizontal lines prior to the vertical sync pulse at the very bottom of the video screen.

Fig. 5.18 is an oscilloscope display showing the relationship between a playback video waveform and the head switching pulse. On a TV monitor with properly adjusted vertical height and linearity, this switching point is below the normal screen viewing area so it is not visible. But it can be observed by adjusting the vertical hold control on the monitor to bring the vertical synchronization pulse up into the display viewing area. You will then notice a very fine line 6½ lines above the vertical sync pulse. This is the time location where the video heads are switched.

A special monitor called a *cross pulse monitor* can display the head switching point 6½ horizontal lines above vertical sync.

Video head switching occurs before the vertical sync pulse to prevent noise interference that could desynchronize the TV vertical circuits causing the picture to roll, become unstable or jump up and down. Designing the switching point placed 6½ lines before vertical sync provides time for the TV horizontal automatic frequency control cir-

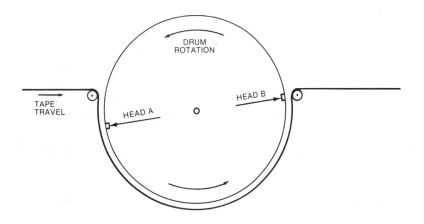

Fig. 5.15. As one video head is lifting away from the tape, the 180 degree opposite head has already made contact.

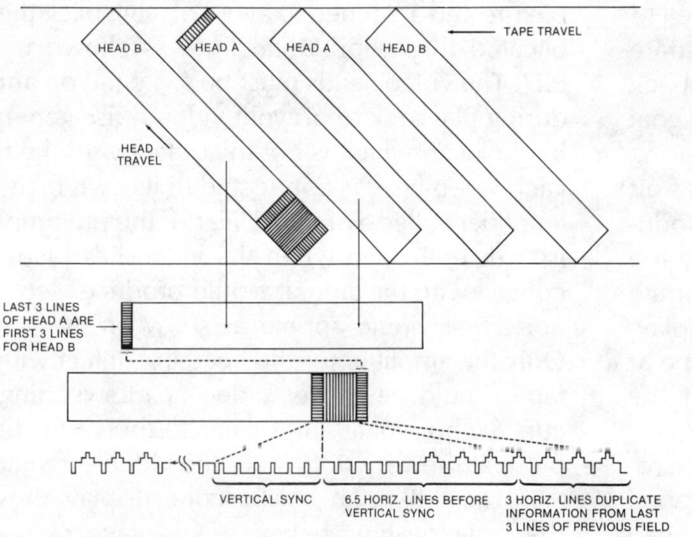

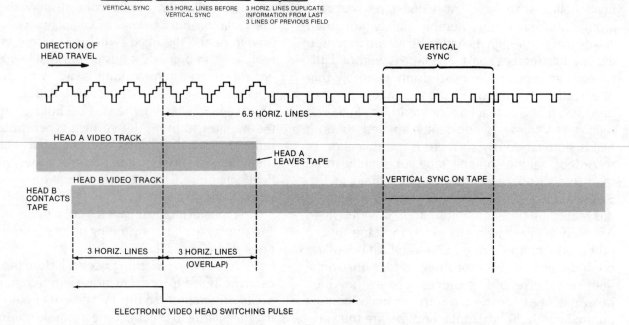

Fig. 5.16. Video heads A and B record the same information for 3 horizontal lines.

Fig. 5.17. The video head switches 6.5 horizontal lines before vertical sync (playback mode).

cuitry to recover from timing error disturbances caused by electronic or mechanical instabilities in the machine. Very precise timing circuitry found in the VCRs servo circuitry controls the position of the switching point. The point is not visible on a properly adjusted video screen. It is important to note that the vertical sync pulse interval must be recorded at exactly the same place on the video head record track with each sweep of the head. See Fig. 5.19.

Guard Bands

Video tape is passed across the video drum assembly at a slight angle so the video information is recorded at an approximate five-degree angle referenced from the edge of the video tape. Some of the very first helical-scan recorders had open reel-to-reel tapes with a space between each of the video record tracks known as guard bands. Video guard bands are thin areas on the tape where no signal is recorded. Guard bands were required to prevent interference on the TV screen caused by information from adjacent recorded tracks being detected by the head during playback. This occurs if head A also detects information from video track B. This crosstalk produces very fine, closely spaced lines of interference on the screen. Video guard bands are spaced to prevent a video head

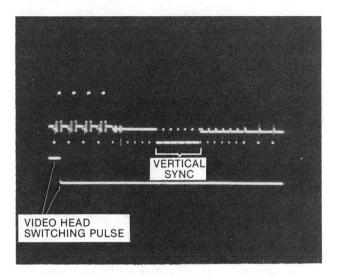

Fig. 5.18. A playback waveform with vertical sync and the video head switching pulse.

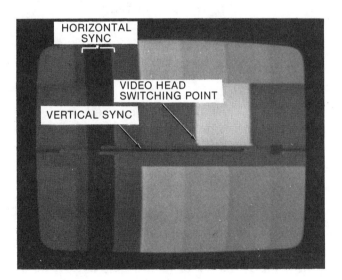

Fig. 5.19. The video head switching point, vertical and horizontal sync are shown on a Cross Pulse Monitor.

from picking up adjacent track information. Crosstalk is also called *cross modulation*.

The use of guard bands represents wasted tape so engineers developed a technique for preventing Y signal crosstalk and eliminating the need for a guard band. This technique has been termed *zero guard band*.

Azimuth Recording

The principle of *azimuth recording* is important in eliminating crosstalk and the need for guard bands. From magnetic tape recording theory, a designer knows that the maximum output signal from a head occurs when exactly one-half the recorded wavelength is within the boundaries of the video head gap such that the gap is exactly 90 degrees to the recorded signal. When an audio tape recorder has high frequency losses from tapes recorded on another machine, yet correctly plays back its own recordings with good high frequency response, the probable cause is the angle or azimuth of the audio head itself. If the audio recorder head gap is not mounted exactly 90 degrees perpendicular to the recorded tape signal as shown in Fig. 5.20, a reduction in high frequency response occurs. It is now no longer possible for exactly one-half the recorded wavelength to fit in the head gap. What occurs instead is a deterioration of high frequency playback signal when magnetic domains of opposite polarity are sensed by the head. This process cancels some of the correct signal field intensity.

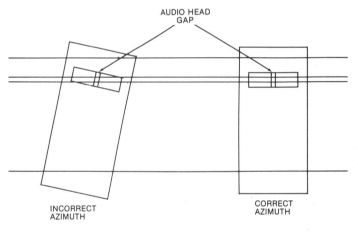

Fig. 5.20. Correct and incorrect audio head azimuth.

By mounting the head so it is perpendicular to the tape information, the video guard band can be eliminated while still providing adequate isolation for the luminance portion of the recorded signal. The video head is constructed with the gap at a slight angle. In VHS machines, one head has a gap placed at an angle of plus six degrees while the angle of gap for the other head is at minus six degrees as shown in Fig. 5.21. The effective azimuth difference between the two heads is 12 degrees. Heads in Beta machines have gap angles of plus and minus seven degrees for a total of 14 degrees azimuth difference.

VHS video head A records a track of informa-

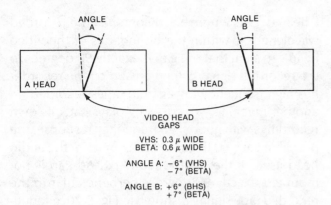

Fig. 5.21. Video head gaps have different azimuths.

tion at a six degree angle to the perpendicular. Then when head A reads that same track during playback, it picks up its own signal and not that of the other track. However, as shown in Fig. 5.22

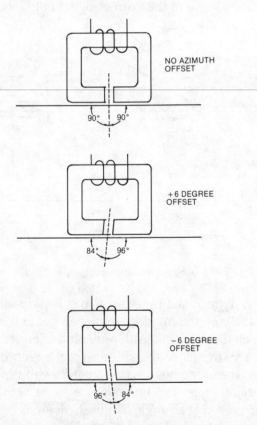

Fig. 5.22. Azimuth offset enables a video head to detect only the signal it recorded.

video head B, mounted 12 degrees away, detects virtually none of the recorded luminence information from track A. Therefore, azimuth recording eliminates the guard band and provides sufficient video track isolation to prevent crosstalk. Neither Beta I nor VHS Long Play (LP) formats have guard bands, but VHS Standard Play (SP) recording does produce a guard band.

Fig. 5.23 describes the use of guard bands in the Panasonic PV1100 operating in the SP mode. Each of the two video head cores are 38 microns wide with a video head gap of three microns. The two video head gaps are designed at plus and minus six degree azimuth offsets. The video record tracks on the tape are equal to the 38 micron width of the video head core material and each video head track has a 20 micron wide guard band in the standard play mode.

Chroma or color information is recorded at a lower frequency so another means must be used for isolating the chroma signals on two recorded tracks. This will be discussed later.

HIGH DENSITY RECORDING

To maximize the use of video tape recording surface while eliminating the need for guard bands, the video record tracks are partially overlapped using a technique called *high density recording*.

Fig. 5.24 shows what happens when the PV1100 begins recording in LP mode. The video head tracks are caused to overlap by nine microns. During the record process, video head A writes information on the tape as the capstan motor advances tape and video head B begins to write its information partially recording over information previously written by video head A. This procedure alternates with head A overwriting a nine-micron-wide strip of track B and vice-versa. The information first recorded by video head A loses some of the effective width when video head B re-records over track A because video head B reorients the magnetic domains in the nine micron strip. During playback, video head A reads its previously recorded track and recognizes enough of its recorded information to faithfully reproduce the original signal, although the width of the video head playback track has been reduced from 38 microns to 29 microns.

Engineers discovered that it was possible to further slow tape speed and use a smaller video

Magnetic Recording Theory   107

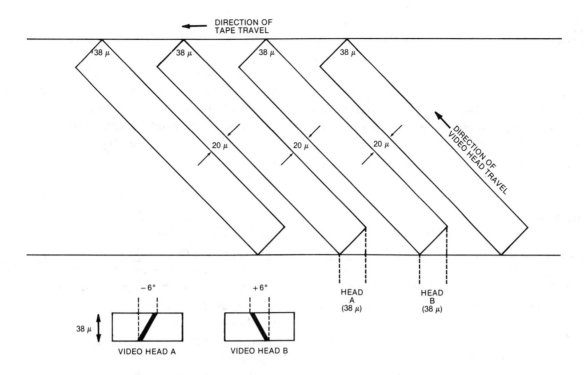

Fig. 5.23. Guard band spacing in SP (Panasonic PV1100).

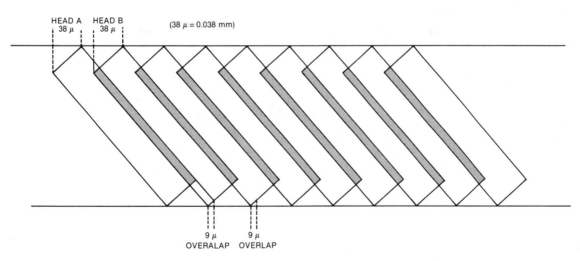

Fig. 5.24. The PV1100 in LP mode has an effective RF envelope width of 29 microns (0.029mm).

head core to pack the video RF tracks even closer. This resulted in the introduction of SLP or EP (Super Long Play or Extended Play) speeds. Early versions of machines capable of these speeds have video heads which are only 30 microns wide using video record overlap technology.

However, a trade-off developed between track width and overlap. Reduction in video head size caused a slight reduction in video picture quality, which prompted the introduction of video tape recorders using 4 heads. Two heads are required for record and playback, so one pair of heads was designated for the SP mode and a second pair was dedicated to LP and EP modes. The different writing speeds meant the use of unequal size head pairs. This caused a crisper video image in SP because wider heads are used during this speed. Using narrower heads in LP and EP was economical because it provided longer record/playback time on the same length of tape.

Video tape recorders with four heads can produce improved special effects in playback because designers could select the video heads which give the best playback picture for slow motion and still frame.

They discovered that by selecting video heads with unequal widths, they could achieve better special effects. Many of the new VCRs have video head cores whose sizes vary slightly. Fig. 5.25 shows the record track for the two-head Panasonic PV5000. The video head sizes in this model differ by five microns. Video head A is 26 microns wide, video head B is 31 microns wide. In SP speed the video heads have two different guard band widths—32 microns and 27 microns.

As shown in Fig. 5.26, operation in LP produces a three micron guard band and a two micron head overlap of the adjacent track.

In SLP mode, as described in Fig. 5.27, no guard band exists and the overlaps alternate between 6.7 and 11.7 microns. However, the effective playback video head track width is 19.3 microns wide for each of the video playback tracks. This is significant because the relationship between the width of the video head core and the effective video playback track directly affects the quality of playback during special effects such as search forward and reverse, still playback and slow motion playback.

Fig. 5.28 describes the relationship between video head A and B and respective recording tracks. The RF envelope produced by the video heads generates a noise free TV image.

Fig. 5.28 also describes the relative angle changes when the video head touches the tape and when the tape is stopped. During still frame playback, the video tape stops and the moving drum assembly causes video head A to begin its excursion up the video head track. It first encounters track B information which it doesn't recognize due to the azimuth difference. The small amount of track A information present at the same point is recognized and becomes the radio frequency (RF) output waveform envelope. As head A

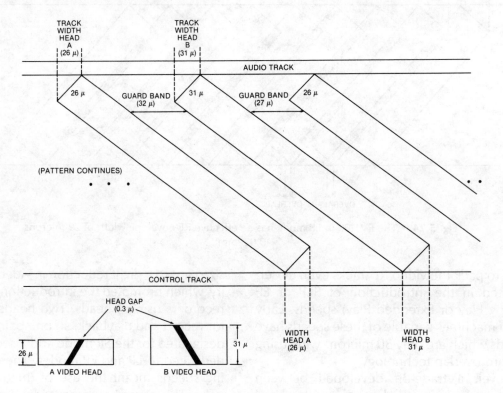

Fig. 5.25. Video head tracks in SP (Panasonic PV5000).

Magnetic Recording Theory 109

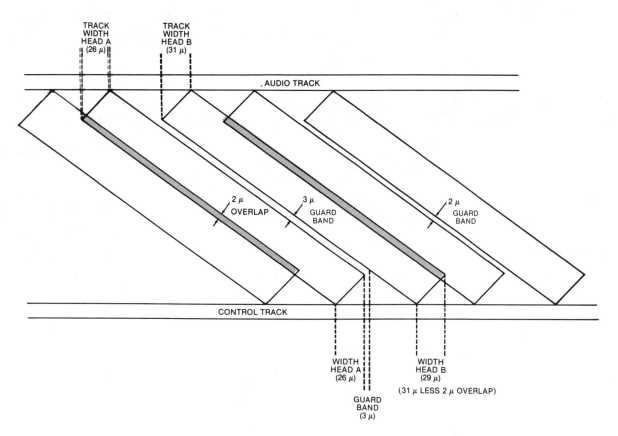

Fig. 5.26. Video head tracks in LP (Panasonic PV5000).

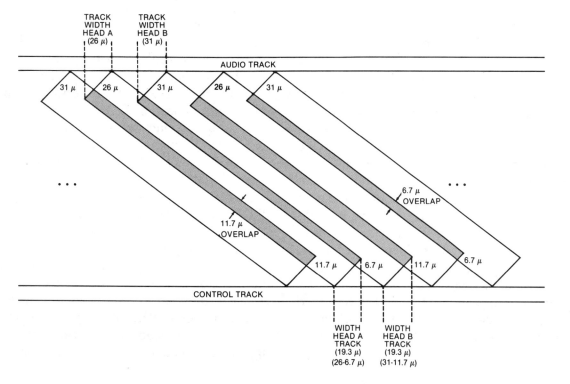

Fig. 5.27. Video head tracks in SLP (Panasonic PV5000).

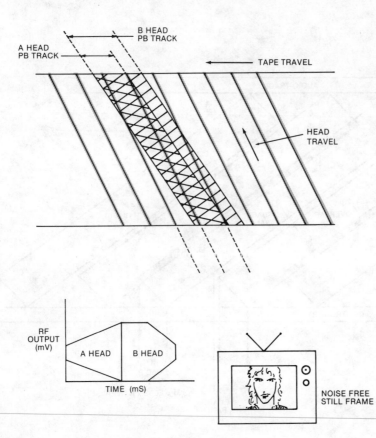

Fig. 5.28. Video output during SLP still frame playback (Panasonic PV5000).

continues up the tape, it detects more of the recorded A track and the RF amplitude increases. At all times during still frame playback, some track A information is detected so the coil in head A always produces some RF output. When head B begins to move up the tape, it covers and detects a large amount of track B record information and its output signal is high. But the RF information begins to deteriorate as the head moves up the track because it detects less track B information and more track A recorded information. Head B is five microns wider than head A so it detects slightly more RF information as it passes over the tape. If the B head weren't five microns wider, it would run out of recorded track B information during a pass and produce no RF output near the last part of its excursion over the tape. No output occurs when a head doesn't detect recorded video information or when the recorded information is traced by a head with the wrong azimuth. The absence of an output signal causes snow on the TV screen. The place where the tape stops is also important in positioning the video heads for proper scanning of the recorded track information and for producing a noise-free still-frame picture. Stopping the tape at the correct position is controlled by the capstan servo and will be discussed in the next chapter.

Table 5.1 compares the video head sizes for a number of Panasonic VCRs.

Early production VCRs couldn't produce noiseless still-frame playback because the video head size was too large to properly scan the recorded RF tracks. The TV screen displays were so poor that designers decided to mute, or blank out the video during playback pause. Fig. 5.29 describes the playback RF envelope and the noisy TV picture during still-frame.

During the search or cue-and-review functions, noise bands are visible on the TV screen. These bands are caused when a video head crosses either a guard band or the recorded track of the opposite head. During search functions, the tape is advanced at approximately 10 times normal

Table 5.1. *Early Version Panasonic VCRs and Their Video Head Sizes*

RECORD/PLAYBACK SPEEDS AVAILABLE	MODEL #	VIDEO HEAD CORE WIDTH	
		Head A	Head B
SP/LP/SLP		30 microns (0.030mm)	30 microns
"	PV1200	"	"
"	PV1210	"	"
"	PV1265	"	"
"	PV1270	"	"
"	PV1275	"	"
"	PV1280	"	"
"	PV1300	"	"
"	PV1310	"	"
"	PV1370	"	"
"	PV1400	"	"
"	PV1470	"	"
"	PV1480	"	"
"	PV1600	"	"
"	PV2600	"	"
"	PV1650	70 microns (0.070mm)	90 microns (0.09mm)
"	PV1750	26 microns (0.026mm)	31 microns (0.031mm)
"	PV1770	"	"
"	PV1780	32&30 microns	45&32 microns
		(4 head unit)	
"	PV3000	26 microns	31 microns
"	PV4000	"	"
"	PV5000	"	"
SP/LP	PV1000	38 microns	38 microns
"	PV1000A	"	"
"	PV1100	"	"
"	PV1500	"	"

speed and the respective read angle with which the video head crosses the tape changes dramatically. In these modes the heads cross several recorded tracks during each scan cycle. As the head crosses a guard band or a recorded track with opposite azimuth angle, the output from the read head is zero resulting in bands of noise or snow across the TV screen. Fig. 5.30 is a picture of snow bands occurring on a current model VCR in search mode. The number of noise bands on the TV screen closely approximates the speed with which the tape is being advanced. For example, if nine bands of snow are counted on the screen, the VCR is advancing at approximately 10 times the normal speed because the head is passing over 10 individual RF tracks for every rotation of the head drum assembly.

Fig. 5.31 is an oscilloscope display of the measured RF envelope and the head switching pulse during search.

The output waveform and TV screen image for forward and reverse search modes are shown in Fig. 5.32 and Fig. 5.33. These figures also show the approximate angles with which the heads strike the tape. In the PV5000 different size video heads reduce visible snow lines on the screen. The 31-micron head produces an effective track width of 19.3 microns in SLP record mode causing head B output at all times in the search mode. This substantially reduces the size of the visible noise bands. When searching on an SP-recorded tape, the guard band becomes quite visible and the noise bands on the screen are much wider.

Beta machines also use varying head widths for special effects, but these units don't produce guard bands in any of their speeds.

112 VCR TROUBLESHOOTING & REPAIR GUIDE

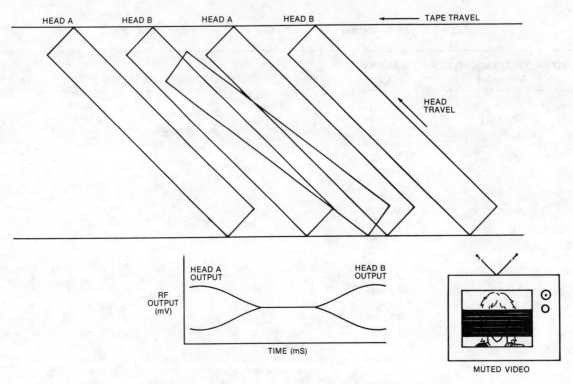

Fig. 5.29. Playback video muted on early VCRs during pause because of poor picture quality.

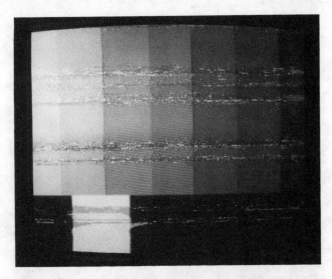

Fig. 5.30. If, during search, the video head crosses an RF track of opposite azimuth or a guard band, snow bands like these occur.

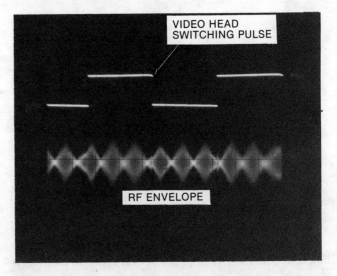

Fig. 5.31. Video head switching pulse and the RF envelope during search.

Fig. 5.34 describes the tape path for a typical VHS machine. The tape for the early Beta machines is shown in Fig. 5.35. Fig. 5.36 defines the tape path for current Beta VCRs. Although the tape paths for early and current Beta machines look opposite, tapes are interchangeable.

Fig. 5.37 is a closeup of the tape path as it travels from left to right past the full track erase head, around the drum and past the audio erase and record heads. The full track erase head removes all previously recorded audio, video and control track information. It uses the bias frequency generated for recording the audio signal as

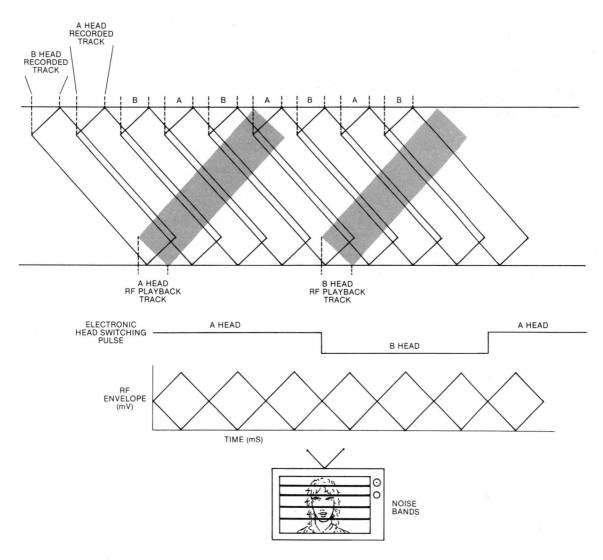

Fig. 5.32. Forward search (cue) SLP speed (Panasonic PV5000).

a high amplitude erase signal and demagnetizes the tape before the tape passes around the video head drum. This head is de-energized during audio or video insert edit modes.

As the tape moves past the full track erase head, it passes through a series of guides that position it over the rotating head drum assembly. The tape is drawn across the head drum assembly at a slight angle to enable helical scan recording. To achieve accuracy, the tape must be carefully guided along the head drum assembly by the entry and exit guides adjacent to the head drum. Proper tape path movement around the scanner (head drum assembly) is controlled by a bump called a *rabbet* located on a small ledge on the lower part of the drum assembly. The rabbet keeps the tape on a precise path around the drum. Many Beta machines have small plastic levers on top of the stationary drum assembly with two plastic arms that gently press on the top edge of the tape to keep the tape placed firmly on the rabbet as it passes around the drum.

Fig. 5.38 is a complete Beta head assembly showing the plastic arms and the guide rabbet placed at a slant around the drum. The tape exits the drum assembly through the exit guide and passes by other guides as it approaches the audio head assembly.

Beta and VHS audio head assemblies have two separate heads as shown in Fig. 5.39. The first head is the *audio erase head*. It is used to delete any previously recorded audio signals and is only

114 VCR TROUBLESHOOTING & REPAIR GUIDE

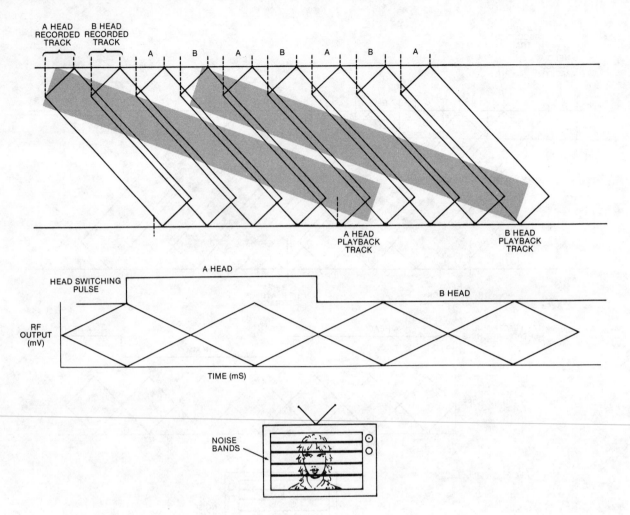

Fig. 5.33. Reverse search (review) SLP speed (Panasonic PV5000).

energized during normal record and audio insert editing. The audio erase head removes the audio signal in the same way that the full track erase head erases the entire tape.

The second head, *audio/control head*, is divided into two head sections—the audio head, and the control track head. The audio head on top is for recording the linear audio signal. It has two record channels if the unit has the stereo linear audio option. Channel one (the left channel) is at the very top edge of the tape; channel two (the right channel) is at the lower part of the audio section of the tape. The lower control track head is used to record the control track pulse on the tape. Control track pulses let the machine identify where to place the video head during playback. The procedure for this operation will be described later.

Fig. 5.40 describes the video tape format for VHS and Beta machines. Fig. 5.41 shows the audio track tape sections produced by a linear track stereo VCR. There are slight differences between the monaural and linear stereo video tape formats.

The effect of the full track erase head can be seen on a TV monitor when a previously recorded tape is rerecorded. For the first few seconds of tape travel, a rainbow pattern is often visible on the TV screen. This rainbow pattern zigzags back and forth across the screen, works its way from the top to the bottom, and eventually disappears. What you see is the remnants of the previous video program. The video head is able to record over and erase most of the previous video but some of the color information remains on the tape because the video record signal is not strong enough to completely realign the polarities of all of the previous magnetic domains. Fig. 5.37 shows a few inches of

Magnetic Recording Theory

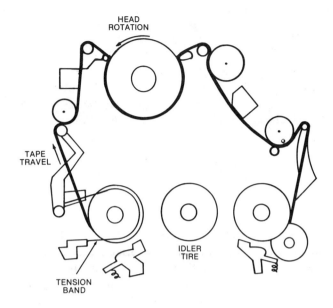

Fig. 5.34. Typical VHS tape path layout (top view).

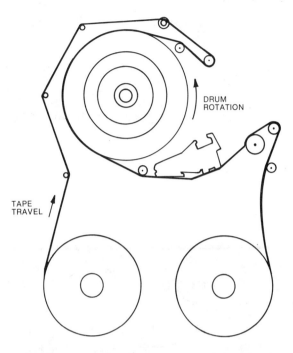

Fig. 5.36. Current Beta tape path layout.

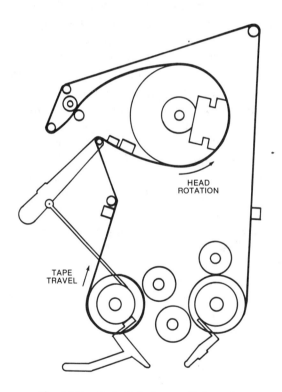

Fig. 5.35. Early Beta tape path layout.

tape between the full track erase head and the video head drum assembly. These few inches of tape cause the rainbow pattern that eventually disappears as the completely erased tape passes the full track erase head and moves across the video head drum assembly. The duration of this rainbow pattern is determined by the speed with which the new recording is made. It takes longer for the completely erased tape to get from the full track erase head to the video heads on the drum in SLP than it does in SP. SLP is 1/3 the speed of SP so the rainbow pattern will appear on the screen three times as long. This is why some people play a few seconds of tape, then start recording. The next recording will ensure that no rainbow pattern is present.

The audio erase head is close in front of the audio record head to prevent the same effect from occurring in audio. If an audio erase head weren't present in the machine, a strip of tape would contain both old and new audio programs. So, to quickly eliminate previous audio programming, an audio erase head was placed in the tape path just prior to the audio record playback head. The audio signal is then recorded using conventional audio record methods.

TIME BASE ERROR

Mechanical VCR instabilities during record and playback cause timing differences or *time base errors* which occur in the horizontal, vertical, and color synchronization pulses. Synchronization

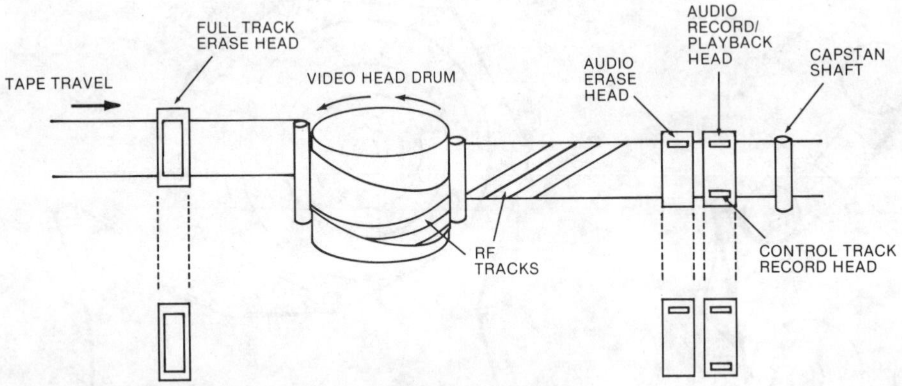

116 VCR TROUBLESHOOTING & REPAIR GUIDE

Fig. 5.37. VHS tape path (front view).

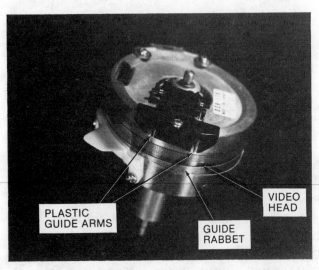

Fig. 5.38. Beta head drum showing critical tape guide components.

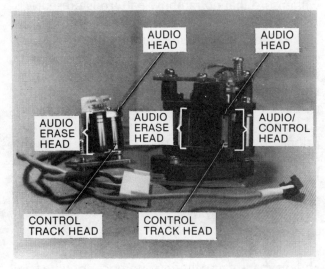

Fig. 5.39. VHS and Beta audio head assemblies.

pulses must occur at precise intervals and the sync pulses must remain for very specific periods of time. Any slight variation in synchronization pulse intervals can cause vertical jumping, horizontal back and forth wavering, or color hue and saturation differences.

One time base error is caused by expansion and contraction of the video tape itself. If the tape expands or contracts after recording, the physical distance between the recorded information on the tape changes causing video synchronization pulses to occur at different times during playback. The result is an unstable TV picture.

Another form of time base error is caused by the tension on the tape as it wraps around the video head drum assembly. VCRs require that the tape be held under a specified tension as it passes by the video head drum assembly. This tension is mechanically controlled by break bands, or, in the case of direct drive reel motors, by the supply reel motor itself. If the video tape is recorded under one holdback tension and then played back on a machine whose hold back tension is different, time base errors result.

Fig. 4.11 described severe time-base error. Fig. 5.42 shows a slight shifting (horizontal time base error) of the horizontal sync pulse at the video head switching point as seen on the monitor. This is caused by a small (few grams) difference in forward hold back tension between record and playback. As the video tape enters the head drum assembly, the servo section ensures that the video head contacts the recorded tape at a prescribed location. As the tape is drawn around the drum, however, the tiny air gap between the video head drum and the tape is reduced substantially. The horizontal synchronization pulses appear

Magnetic Recording Theory 117

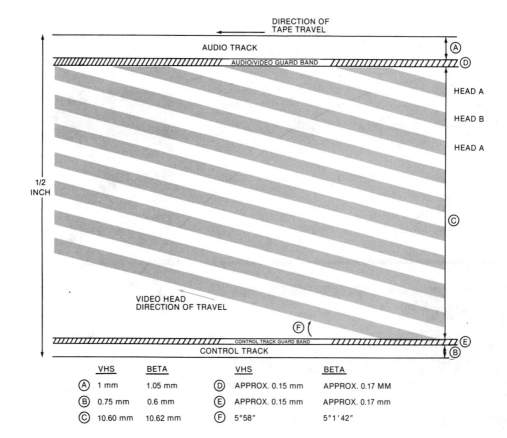

Fig. 5.40. Tape format (VHS SP, Beta BI).

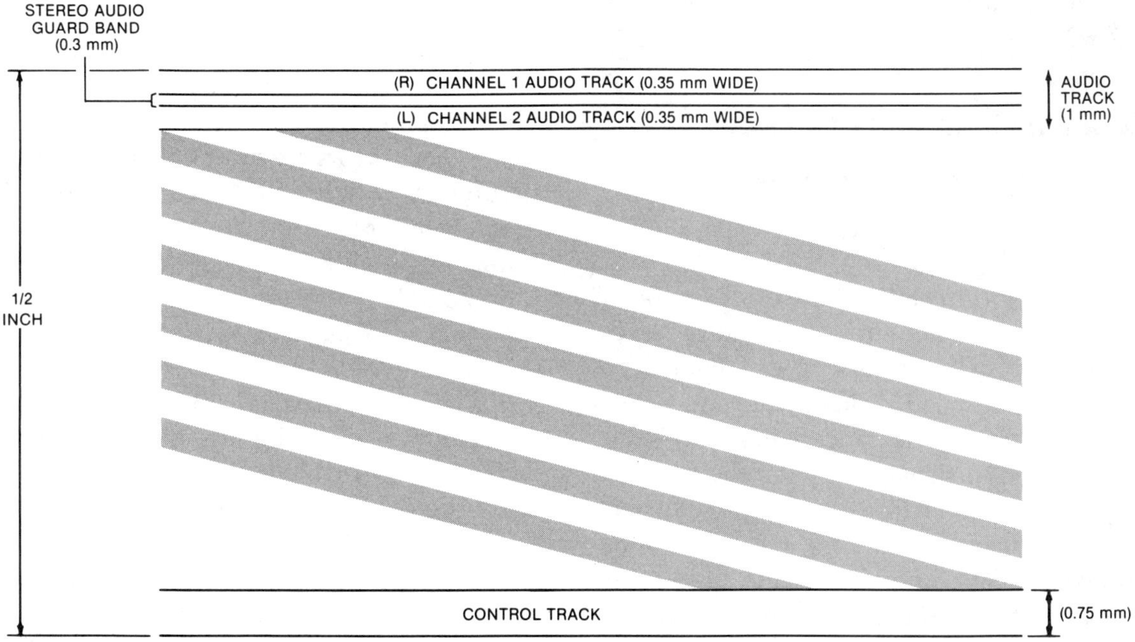

Fig. 5.41. In a VHS recorder with stereo linear audio, the audio/video guard band is eliminated and a 0.3mm space is used as a stereo audio guard band to separate right and left audio channels.

stretched and the monitor attempts to compensate by using its horizontal automatic frequency control circuit. As the video head exits the tape, the horizontal synchronization pulses reach a point where they are most separated. At this time, however, the next head begins to contact the video tape, and the horizontal pulses for its video field appear closer together so the process repeats. The TV monitor receiver must make dramatic jumps between horizontal time base errors and rapidly compensate for each new horizontal pulse frequency being generated. The result is a bending or tearing of vertical objects in the upper portion of the TV picture.

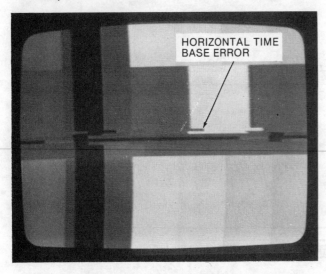

Fig. 5.42. Horizontal time base error caused by minor forward holdback tension differences between record and playback.

Yet another form of time base error can occur as a result of misalignment of the video heads. Video heads must be positioned exactly 180 degrees apart. If one head is slightly advanced relative to the other, the advanced head will start playing back video signals prematurely. This produces synchronization signals which occur early causing the synchronization and picture information to take a slight jump forward in time at the video head switching point. As soon as the second head begins to contact the tape, the video signal and horizontal synchronization impulses jump backward in time. This form of time base error is called *dihedral error* and can be corrected by re-aligning the heads. Dihedral error produces a horizontal jumping back and forth at the video head switching point between the two vertical fields as shown in Fig. 5.43.

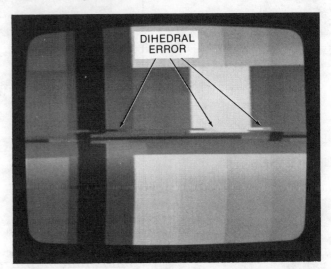

Fig. 5.43. Dihedral error causes horizontal time base error. (Compare with Fig. 5.42 which has no dihedral error.)

Time base errors can also be caused by friction between the tape and the guides in the video tape path and by mechanical vibrations of the video tape itself. High humidity can cause the tape to adhere to the metal guides and drum assembly. Other time base errors can be caused by instabilities in the video tape speed as it is drawn through the machine.

These errors are most noticeable in the higher frequency synchronization pulses including the horizontal and chroma sync pulses. The low frequency sync pulse (vertical rate) is not as affected. Horizontal time base errors generally appear as bending or flagging of vertical objects on the upper portion of the video screen.

The TV horizontal automatic frequency control circuit plays an important part in reducing the effects of horizontal time base errors. Until a few years ago, American-made TV sets operated with extremely stable TV broadcast signals. The horizontal automatic frequency control (HAFC) circuits were deliberately designed to slowly compensate for timing errors and avoid the time base errors caused by external random impulse noise. Such noise could be caused by a passing vehicle or a nearby electric motor. With the introduction of VCRs, American TVs needed HAFC circuits that provided faster compensation time. Because older TVs are not able to respond quickly enough to the horizontal time base errors introduced from the VCR, a VCR connected to an older TV may cause severe bending or flagging at the top of the picture yet appear to operate normally on a newer TV set. Fig. 4.11 in Chapter 4 shows flagging.

Consumer video tape recorders don't record color subcarrier frequencies directly on the tape. Only very expensive broadcast quality helical-scan VCRs can directly record color frequencies. Color recording in consumer machines is done by a process called *color under*, an economical way to record chroma signals and still eliminate the visual effects from time base errors.

Color or chroma information is broadcast by TV stations using a subcarrier frequency of 3.579545 MHz (3.58 MHz). The higher the sync frequency the more sensitive the synchronization circuits become to minor time base errors. Consumer VCRs don't record chroma information onto the tape at 3.58 MHz. Instead the chroma subcarrier is converted to a frequency below that for the Y FM information, thus the term *color under*. The color-under process was developed to reproduce color TV programs while minimizing time base errors produced by the VCR during record and playback. Chroma time base errors are seen on the video screen as hue (tint) changes, and in severe instances, as loss of chroma synchronization or loss of all color. Loss of chroma synchronization can cause either a rainbow effect or a total loss of color.

Knowing how TV signals are broadcast helps you understand the color-under process. Video information on the TV screen has a band width of approximately 4.2 MHz as shown in Fig. 5.44.

Picture detail is contained in the black and white portion of the video signal. This is called *luminance* and is designated Y. The Y signal varies in intensity from black through shades of gray to 100 percent white and contains all of the fine detail in the picture scene as well as the vertical and horizontal sync information. Fig. 5.45 is an oscilloscope display of the Y signal without the presence of color information.

Color information is modulated using a 3.58 MHz subcarrier and "painted" on the black and white detail in the video picture. Fig. 5.46 shows how chroma is "painted" over the black and white detail so the resultant broadcast video contains both Y and chroma information.

Fig. 5.47 is a broadcast video signal that contains brightness detail, burst, chroma, and horizontal sync information.

Colors are represented by different phases of the 3.58 MHz subcarrier as referenced by a color sync signal called "burst." The color subcarrier signal consists of side bands of approximately 600 kHz that extend above and below the main 3.58 MHz subcarrier frequency. A frequency of 3.58 MHz is selected low enough to allow the 600 kHz side band to fall within the 4.2 MHz band width upper limitation and yet retain the color subcarrier frequency in the upper portion of the video frequencies. The 3.58 MHz chroma subcarrier frequency also reduces the visual effects on the TV screen caused by subcarrier frequency modulation.

The audio in the TV broadcast signal is modulated around a 4.5 MHz carrier frequency. The close proximity of the color and audio subcarriers required that the color subcarrier frequency be selected to minimize interference with the audio carrier. The effects of the chroma subcarrier frequency appear as small dots and are less noticeable on the TV screen because the 3.58 MHz color information has a fixed relationship with the hori-

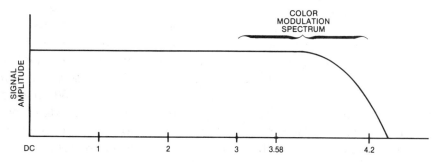

Fig. 5.44. The video frequency band width.

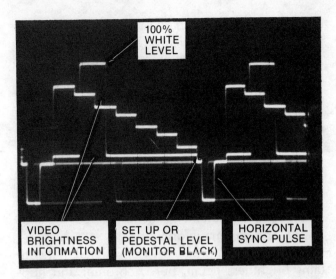

Fig. 5.45. Video (Y) signal only (no color information).

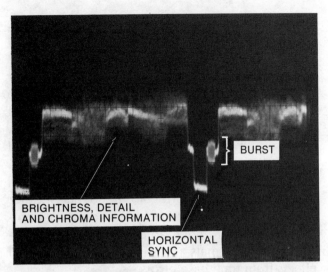

Fig. 5.47. A broadcast video signal containing brightness, detail, burst, chroma and horizontal sync information.

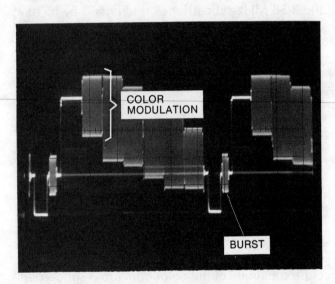

Fig. 5.46. Video (Y) and color information.

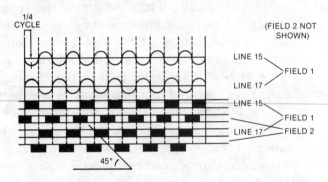

Fig. 5.48. Color modulation. Adjacent even or odd lines have opposite subcarrier phase relationship.

zontal picture line rate (15,734 KHz). The subcarrier is not an even multiple of the horizontal frequency and appears opposite in phase on alternating lines in the same field on the TV screen. The screen dot pattern from the chroma subcarrier is less noticeable because it occurs at a 45-degree diagonal on alternating lines. Fig. 5.48 shows the relationship between the chroma subcarrier frequency and the dot pattern produced on the TV screen.

Notice that the phase of the subcarrier frequency is not flipped on adjacent lines but instead is a continuous oscillation. Each line of color information contains a series of full cycles minus a half cycle so alternating lines appear to be 180 degrees out of phase. Fig. 5.49 is the dot pattern display on a high resolution monochrome monitor screen.

In VHS and Beta machines, the color subcarrier and side bands are converted to 688 kHz for Betamax and 629 kHz for VHS. To record color information at the lower end of the frequency spectrum, some of the lower FM sideband is given up as shown in Fig. 5.50. This produces some loss in black and white detail. The extreme lower portion of the lower FM sideband is electronically chopped via filter networks. VCRs can automatically sense when a program being recorded is not in color. When black and white (monochrome) signals are detected, the filters that cut off the lower part of the lower FM sideband are bypassed to allow full lower side band operation. This pro-

Fig. 5.49. Color subcarrier dot pattern on a TV screen.

vides greater picture detail when recording black and white TV programs.

The color-under frequencies of 629 and 688 kHz were selected to allow enough room for the upper and lower sidebands of approximately 500 kHz for the color under frequencies.

Color-under frequency conversion is accomplished by heterodyning two signals of different frequencies to form resultant frequencies that are the sum of the two base frequencies, the difference of the two base frequencies and the original base frequencies. The color-under frequencies in VHS and Beta are derived by adding the standard 3.58 MHz color subcarrier frequency to a stable frequency generated inside the VCR as shown in Fig. 5.51. For Betamax units this frequency is approximately 4.27 MHz. VHS units use a frequency of approximately 4.21 MHz.

Color information is separated from the video luminance information, as shown by the oscilloscope photograph in Fig. 5.52. Next it passes through bandpass filters and combines with locally generated frequencies. In Betamax, the 4.27 MHz local frequency is combined with the 3.58 chroma frequency to produce 7.85 MHz, 3.58 MHz, 4.27 MHz and 688 kHz signals.

Fig. 5.53 is an oscilloscope display of the 688 kHz and 3.58 MHz separated chroma signal before and after frequency conversion. A low pass filter passes only the 688 kHz signal to a mixing circuit where it is combined with the frequency modulated Y signal. The 688 kHz color signal is then directly recorded onto the tape by the video

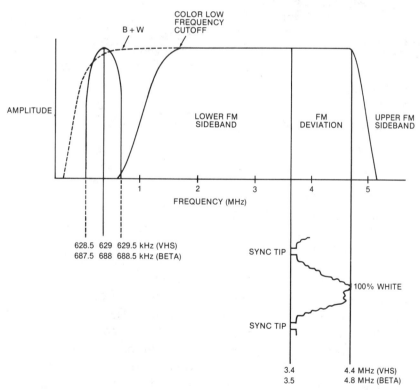

Fig. 5.50. Frequency allocations for VHS and standard Beta video cassette recorders.

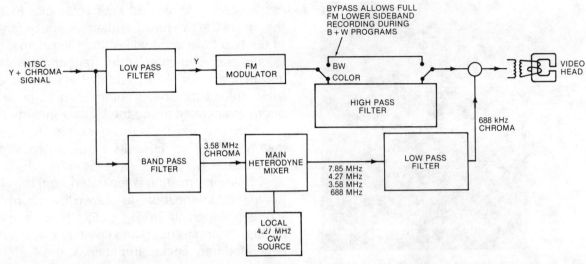

Fig. 5.51. Basic Betamax color-under record circuitry.

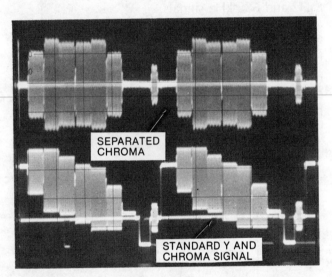

Fig. 5.52. Chroma is separated from luminance for color processing.

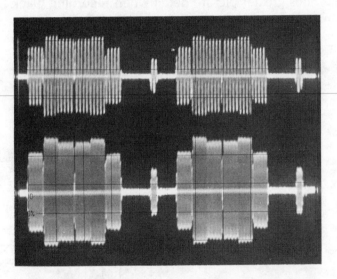

Fig. 5.53. The 688 kHz color-under chroma (top) compared with the original 3.58 MHz signal (bottom).

heads. The combined FM Y and 688 kHz color signals are then directly recorded on the tape through the video heads. The color-under chroma signal is recorded on the tape using the same principles as that for audio recording. Both audio and color signals require AC bias during recording to avoid distortion. The Y FM signal becomes the bias during recording of the color-under signal.

During playback, the color and luminance FM signals are read off the tape, separated by frequency and processed individually. The color-under signal contains time base instabilities created as a result of the record and playback process.

Fig. 5.54 shows how the 688 kHz chroma signal is converted to the standard 3.58 MHz signal and how time base errors are substantially minimized.

Playback Y and color-under information extracted off of the tape by the video heads enters the circuitry from the upper left hand corner of the diagram. The Y signal is applied to the FM demodulator where it is reconverted into the normal luminance waveform containing brightness, detail and sync information. The 688 kHz chroma signal is separated in a low pass filter which eliminates the high frequency Y signal and allows only

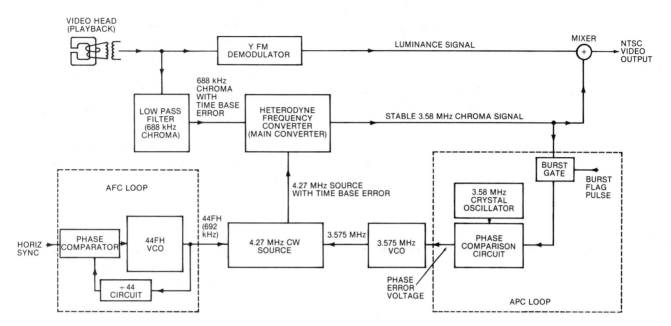

Fig. 5.54. In early Betamax units, chroma frequency and phase control circuits eliminate nearly all time base errors.

the lower 688 kHz chroma signal to pass. This 688 kHz signal contains time base instabilities.

The color-under signal containing time base errors is mixed with a 4.27 MHz continuous wave (CW) signal which also contains time base errors to produce a stable 3.58 MHz output. If the time base errors caused a slight increase in the 688 kHz signal, the 4.27 MHz signal, containing the same time base error, would increase proportionally. The frequency difference output from the mixer is maintained at 3.58 MHz by mixing the two varying input signals. This stable mixer output can only be achieved when the two frequencies change in unison.

The lower left of Fig. 5.54 shows a feedback loop for automatic frequency control (AFC). The input to the AFC loop is a horizontal sync signal from the demodulated video off the tape. These horizontal sync pulses are the same pulses which provide horizontal synchronization for the TV monitor. They also contain time base errors from the tape. The AFC loop circuitry contains a voltage controlled oscillator (VCO) which operates 44 times that of the horizontal sync and is identified as 44fH (692 kHz). The output of the 44fH VCO is input to a divide-by-44 circuit. The output of the divide-by-44 circuit is passed to a phase comparator which compares the incoming horizontal sync from the demodulated video with that of the internal 44fH VCO. The phase comparator recognizes time base errors and generates a DC correction voltage for the 44fH VCO. The 692 kHz output frequency from the AFC loop contains the horizontal time base errors. It is interesting to note that the proportional time base error generated in the 692 kHz signal is near to that which is found in the 688 kHz signal because the time base errors are multiplied 44 times. The 688 kHz color-under frequency is actually 43.75 times more than the horizontal rate so the end result is that the AFC loop tends to overcorrect by a factor of .25fH. The slight difference still exists in minor chroma phase errors but these errors are compensated in another circuit.

The 692 kHz signal containing the time base errors is mixed with a 3.575 MHz signal to produce a 4.27 MHz output containing time base errors. This signal is mixed with 688 kHz in the main converter. The AFC loop has a wide dynamic range for correction of time base errors and is responsible for most chroma time base error corrections.

The 688 kHz signal is only 43.75 times that of the horizontal frequency. Since the nearest whole number to 43.75 is 44, the 44fH frequency was chosen. The difference between 44 and 43.75 requires minor phase correction which is achieved by the automatic phase control circuitry or APC. Fig. 5.55 describes how this is accomplished. The 3.58 MHz chroma signal from the main converter is sampled by the APC circuitry where it encounters a burst gate circuit which is switched on and off to allow color signals to pass only during specific time periods. Switching is controlled by a horizontal synchronization burst flag pulse that is delayed slightly to match the position of the burst pulse in the chroma signal. Only the burst pulse is allowed to pass through the burst gate.

Burst consists of 8–10 cycles of 3.58 MHz oscillation and is the color sync signal broadcast from the TV station to synchronize the chroma circuits in the receiver. Burst is passed to a phase comparison circuit where it is compared with the output from a very stable 3.58 MHz crystal oscillator. The phase comparison circuit recognizes small phase variations in the playback burst signal and produces a phase error correction voltage for the 3.575 MHz voltage controlled oscillator. The 3.575 MHz VCO contains slight chroma phase error correction signals which are passed to the 4.27 MHz CW mixing circuit. The 3.575 MHz VCO output containing phase error corrections is mixed with the 692 kHz signal containing frequency corrections. Frequency correction acts as a coarse adjustment and phase correction acts as a fine adjustment. Both corrections are passed to the main frequency conversion circuit to produce a very stable 3.58 MHz output chroma signal.

VHS systems use the same basic procedure to correct for time base errors, so the same basic block diagrams could be used for VHS circuits although some of the frequencies are different. For example, the VHS AFC loop uses a frequency of 40 times fH instead of 44 fH. The color-under subcarrier frequency is 629 kHz (actually 629.371 kHz) and the VCO controlled by the APC circuit is closer to 3.58 MHz instead of 3.57 MHz. There are other differences in the way Beta and VHS systems accomplish chroma processing. These involve the electronic circuitry for eliminating crosstalk. But this basic block diagram is sufficient for understanding the elimination process for the time base error.

After studying the APC block diagram in Fig. 5.54, one might wonder if it wouldn't be simpler to have the phase correction occur at 688 kHz since it would eliminate one of the mixers in the circuit. This is not done because the 688 kHz color-under burst contains two cycles, and this is not enough to perform phase comparison. Normal burst at 3.58 MHz contains between eight and ten cycles of oscillation.

Chroma Crosstalk Elimination

Previous sections described how Y FM signal crosstalk can be eliminated by the use of azimuth recording techniques. Crosstalk in a luminence signal can be eliminated as shown in Fig. 5.55 by recording the signal using high frequency FM for the Y portion.

Azimuth recording techniques work effectively only for high frequency signals but have little effect on crosstalk elimination for frequencies below 1.5 MHz because some adjacent track information is detected by the opposite head as shown in Fig. 5.56. Therefore, when color under is used, azimuth recording techniques provide little isolation for the color portion in adjacent channels.

Complex electronic circuitry is used to eliminate chroma crosstalk on adjacent RF tracks, and this is one area where Beta and VHS formats differ greatly. The end result is the same. Chroma information on adjacent tracks is recorded and played back in such a way that crosstalk is eliminated. Chroma crosstalk cancellation for both tape formats uses the fact that the chroma subcarrier signal at any given point on two sequential horizontal lines in the same field are opposite in phase. Refer-

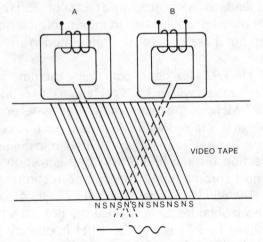

Fig. 5.55. Azimuth recording techniques prevent head B from detecting the head A Y FM signal.

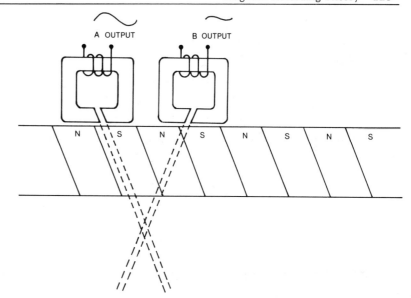

Fig. 5.56. Head B detects more N than S, but since N isn't cancelled by an equal amount of S, some A signal is detected by head B.

ring to Fig. 5.48, note that line number 15 and line number 17 are sequential horizontal lines in field number one, and in the lower portion of the figure line number 15 and line number 17 are the resultant screen dot pattern. Also note that the dots and the subcarrier waveform are exactly 180 degrees out of phase.

Chroma crosstalk cancellation circuitry for both tape formats electronically modifies this phase relationship during record. During playback the original chroma phase relationship is restored. Although the two formats differ in the method used to modify the subcarrier phase relationship, a special circuit called a *comb filter* is the common link between the two processes.

Comb Filter

A comb filter acts like an ordinary hair comb in that unwanted portions of a signal are combed out of the information passing through the circuit. Consisting of a glass delay line and a resistor network, the comb filter shown in Fig. 5.57 has a built-in time delay of one horizontal line (63.5 microseconds). The delay line is caused by the time it takes for a signal to pass through a measured length of glass. Two transducers are attached at each end of the glass delay line. A transducer can be a speaker or a microphone depending on whether it is sending or receiving a signal. The delay line input transducer receives the electronic signal and converts it to acoustical waves which travel through the glass reflecting off the glass edges. When the acoustical waves reach the output transducer, they are converted back into an electronic signal by the output transducer as shown in the figure.

Comb filter applications in TV design have gained popularity recently because of their ability to increase resolution. The TV video band width extends up to about 4.2 MHz. Broadcast color information occupies the space between about 3.2 and 4.2 MHz. Previously it was difficult for older TVs to separate the chroma subcarrier information and the video detail information in this common frequency band. Traps were used to remove or filter the color signal from the Y information, but this resulted in the loss of luminence information above 3.2 MHz and a loss of detail on the TV screen.

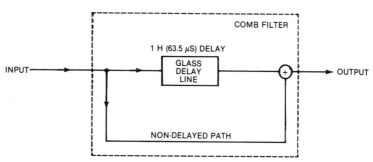

Fig. 5.57. The comb filter uses a glass delay line and a summation to cancel out crosstalk.

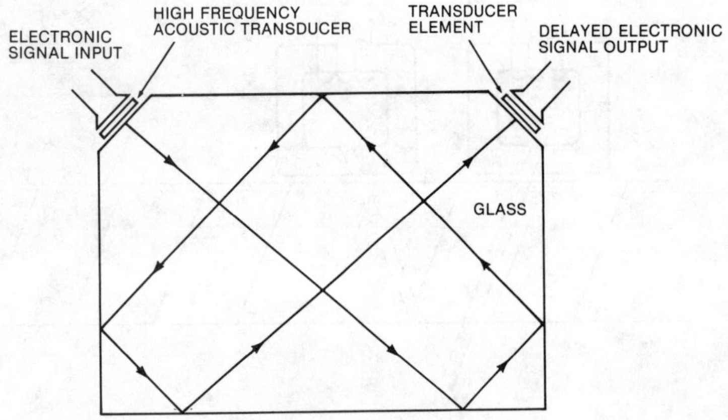

Fig. 5.58. A 63.5 microsecond glass delay line in the comb filter.

A color TV purposely does not display the color subcarrier frequency oscillation on the face of the picture tube because the fine dot pattern it produces is objectionable. The subcarrier dot pattern on the monitor screen in Fig. 5.49 is from a monochrome high resolution monitor. This pattern would not be as noticeable on a color monitor. Unfortunately, the Y signal resolution also suffered as a result of the removal of this pattern. Earlier TVs had a video band width of about 3.1 MHz. This equals approximately 250 lines of horizontal resolution.

The comb filter uses a technique called *frequency interlace* to cause color and Y information to occupy the same frequency band. Frequency interlace causes chroma side band information to fall between the harmonics of the horizontal signal containing video information up to the 4.2 MHz video band limit. The subcarrier frequency is an odd multiple of the horizontal scan rate so the dot pattern created by the subcarrier oscillation is less visible on the TV screen. The comb filter uses frequency interlace in a TV receiver when separating chroma and Y signals. Each sequential horizontal line in the same vertical field has chroma subcarrier information exactly 180 degrees out of phase from the same points on the previous line to reduce the visual effects of the chroma subcarrier on the TV screen. The comb filter adds to or subtracts from the signal on one line and the information contained on the previous line. Since color signals are 180 degrees out of phase they add and cancel out of the video portion of the signal. Y information doubles in amplitude in the comb filter. Reversing the leads on the output of a comb filter causes the opposite effect. The luminence signals cancel because they are out of phase with each other while the color signals are added together and nearly doubled. The same effect also occurs with the harmonics and side bands of the main signal. A comb filter combs the desired information out of a composite video signal improving video resolution while still separating and processing chroma side band information.

Comb filters are also important for eliminating chroma crosstalk which occurs as a result of zero guard bands. Before the comb filter circuitry can function, the color-under chroma signal phase relationship must be modified during the record and playback process. Since the Beta system uses a simpler approach to modify the subcarrier phase relationship, it will be described first.

The Betamax system alters the subcarrier phase relationship on only one of the vertical fields during the record process. This phase modification occurs only when the A head is touching the tape. Phase modification is accomplished by the circuitry between the 4.27 MHz CW source and the main heterodyne mixer. Fig. 5.51 grossly generalizes the local 4.27 MHz CW source block to simplify earlier explanations.

CW Source

Fig. 5.59 shows a detailed block diagram of how the 4.27 MHz CW source is derived. Horizontal sync that has been separated from the incoming video source enters the AFC loop, goes through the phase comparator, and is applied to the 44 fH VCO. A sample of this horizontal frequency multiplied 44 times is passed through the divide by 44 circuitry to produce a horizontal pulse that is fed back to the phase comparator. The phase compar-

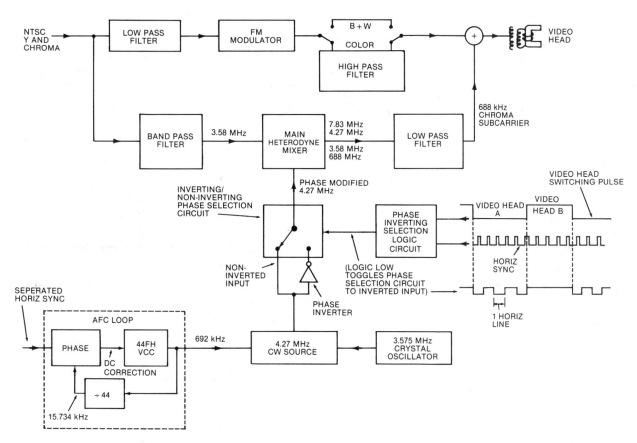

Fig. 5.59. Betamax color-under record circuitry with chroma subcarrier phase modification.

ator generates a DC correction voltage which controls the 44 fH VCO so its output 692 kHz (44fH) frequency is phase-locked to the incoming video signal. The VCO output enters the 4.27 MHz heterodyne submixer circuit and combines with a very stable 3.57 MHz from the crystal controlled oscillator. The 3.57 MHz and 692 kHz signals are mixed together to produce 4.27 MHz, the sum of the two frequencies. The 4.27 MHz output becomes one input to the phase inverter selection circuit. The same signal is inverted and becomes the second input to the circuit. During the time that video head A is recording, the chroma phase selection circuit is receiving horizontal synchronization pulses. These pulses cause a switch to toggle between the inverted and noninverted 4.27 MHz oscillator inputs. The output is a 4.27 MHz CW signal that is phase inverted every other horizontal line. While head B is recording, the phase inverter selection circuit selects the noninverted input causing the 3.58 MHz chroma subcarrier signal to be converted to 688 kHz in the main heterodyne mixer circuit with a modified subcarrier phase relationship. The horizontal lines recorded with head A have a chroma subcarrier phase relationship exactly in phase with each sequential line. The subcarrier phase relationship recorded with head B maintains the original phase alternating relationship.

The 4.27 MHz CW signal is phase locked to the incoming horizontal sync pulses. The 3.57 MHz crystal controlled oscillator is 3.58 MHz, the subcarrier chroma frequency minus ¼ that of the horizontal frequency. This means that the color-under chroma signal is always phase-locked to the incoming horizontal sync multiplied by 44 minus ¼ fH, and the 688 kHz color-under signal maintains the exact phase relationship with horizontal sync as the original 3.58 MHz subcarrier frequency.

The phase inverter selection circuit is driven by a logic circuit which has two inputs. One input is a 30 hertz (actually 29.97 Hz) video head switching pulse derived from the video head itself. This

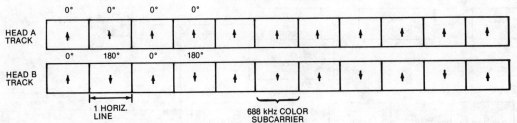

Fig. 5.60. Phase relationship of adjacent RF tracks when 688 kHz color subcarrier is modified.

signal will be described in the servo section. The other logic circuit input is a sequence of horizontal synchronization pulses that cause the phase inverter selection circuitry to alternate between inverting and non-inverting inputs. Thus, in the field that head A has recorded, the phase of every other line is inverted. The resultant output is such that each sequential horizontal line of the A field has an in-phase subcarrier. During the time that head B is recording, the 30 Hz head switching pulse forces the phase inverter selection circuit to remain in the noninverting input which allows the subcarrier phase to alternate during each sequential line. Fig. 5.60 shows the subcarrier phase relationship for each horizontal line of the RF track for heads A and B.

During playback the A head subcarrier phase is restored to its original alternating phase. Playback phase restoration is done by the same circuitry that performed the modification during record. The only difference is that the subcarrier phase now contains time base errors and the crosstalk component from the adjacent track. Fig. 5.61 shows the subcarrier phase relationship to the recorded signal and the restored playback subcarrier during playback. Section A of the figure compares the head A and head B recorded tracks. The color-under subcarrier phase is modified for head A so each sequential line for the head A field is in phase. The head B color-under subcarrier phase isn't modified so each horizontal line alternates in phase.

Fig. 5.61B describes the playback phase relationship of the color-under signal from the video heads. The head A field contains the 688 kHz signal whose phase is exactly as recorded plus the adjacent head B track crosstalk chroma information whose phase is alternating. Crosstalk information is indicated by the dotted arrows. The head B chroma information contains the phase alternating

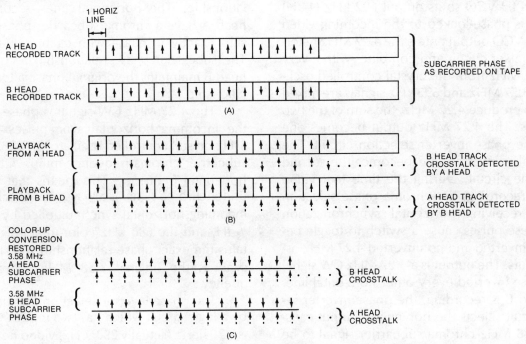

Fig. 5.61. Chroma subcarrier phase relationships during playback.

688 kHz color-under subcarrier and the chroma crosstalk from the A head track whose phase does not alternate. Crosstalk information is again indicated by dotted arrows. Section C shows what happens to the subcarrier phase relationship when the color-under signal is restored to its original alternating phase condition. Only head A information was modified during record so in playback only head A information must be restored.

The main subcarrier of the chroma signal and the crosstalk information are converted so the crosstalk information picked up by head A is converted and the subcarrier phase relationship between each adjacent line is in phase. During playback, head B information does not require modification so its information is not phase inverted. Crosstalk information coming from head B has the same phase relationship for each horizontal line because that is how it was recorded originally. The end result of this process is that the crosstalk information for the A and B heads has the same phase relationship at all times.

The head A and B color-under signals and their crosstalk components are converted back to the 3.58 MHz color subcarrier frequency with its original phase relationship. The crosstalk component from each of the heads is also converted to 3.58 MHz, but each adjacent horizontal line is of the same phase. Once the signal has been converted to 3.58 MHz it is passed to the comb filter circuitry.

The comb filter circuitry differentiates between the main alternating subcarrier frequencies and the nonalternating subcarrier frequencies to identify the crosstalk component and comb them out of the chroma information. Fig. 5.62A describes how the comb filter circuitry actually cancels out crosstalk information. The 3.58 MHz up-converted chroma information enters the comb filter circuitry through amplifier Q1. The chroma information takes two separate paths. The first path is through the nondelayed line and the resistor network where R1 and R2 join as indicated by the smaller solid arrows. Electrical current separates and flows in two directions. Half the current flows through R1; the other half flows through R2. Both legs of this current path eventually reach electrical ground through resistors R3, R4 and capacitor C1. Fig. 5.62B represents the chroma phase relationship between the delayed and nondelayed input signals.

The second chroma information path from

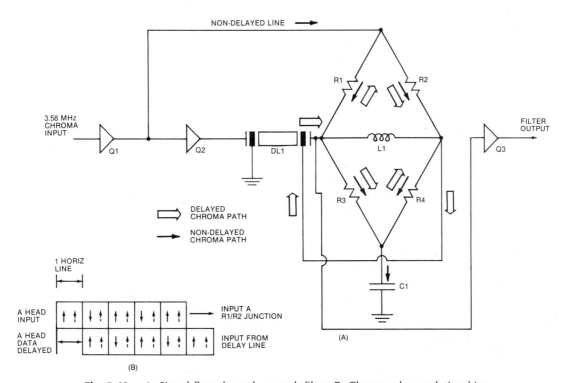

Fig. 5.62. A. Signal flow through a comb filter. B. Chroma phase relationships.

Q1 is through the delay line. The incoming 3.58 MHz up-converted chroma from Q1 enters the delay line through amplifier Q2 which further enhances the incoming signal to ensure that enough signal amplitude remains after passing through the attenuation of delay line DL1. As the chroma signal passes through the comb filter's delay line, it is converted into acoustical vibrations which travel through the glass and are delayed by 63.5 microseconds before entering the resistor network at the junction of R1 and R3. This signal flow is indicated by the larger, hollow arrows. The signal current flows through resistors R1/R2 and R3/R4 to the junction at the right where it is returned to the delay line.

Chroma crosstalk cancellation occurs inside the resistor network when two currents of the same phase and amplitude meet coming from opposite directions. From electronic theory, we understand that if two opposite voltages of equal amplitude are applied across a resistive load, the cumulative voltage drop is zero and no current flows through this load. This is how undesired signals cancel in the comb filter. The nondelayed crosstalk information flows in one direction while the delayed signal from the previous line flows in the opposite direction. Out of phase information is flowing through the resistor network in the same manner. Crosstalk is identified by the dotted arrows. This is what we want to eliminate. The desired signal is out of phase. The reason delay DL1 is in the design is to cause the signal and delayed signal to meet at the R1/R3 junction where the crosstalk is cancelled and the desired signal is approximately doubled.

Since the desired chroma information is phase-inverted at every line, the nondelayed subcarrier signal flowing in one direction is 180 degrees out-of-phase with the delayed subcarrier information flowing in the opposite direction. These two out-of-phase signals add and nearly double in this resistor network. The end result is crosstalk elimination and a doubling of the desired chroma signal to amplifier Q3. Amplifier Q3 boosts the delay-line-attenuated signal to a level sufficient for filter output.

The comb filter requires that color subcarrier information from the two sequential lines has chroma signals of the same amplitude for both horizontal lines. Another important point is that the comb filter requires two lines of information for cancellation. Chroma crosstalk is therefore not cancelled on the first and last lines of the TV screen. This has little effect for two reasons. First, horizontal lines are extremely fine on a TV screen since 525 lines make up the entire picture. Secondly, the first and last lines of video information are respectively above and below the normal viewing area.

While proper operation of the comb filter requires that color crosstalk information is equal in amplitude between any two lines of a given field, this is not always the case. Some crosstalk interference may occur because color information isn't always identical. Crosstalk, then, is substantially reduced but not completely cancelled.

Comb filters also do not completely double the color amplitude signal. Some minor distortion occurs as the chroma signal passes through the glass delay line. This allows some of the adjacent channel crosstalk to pass through the comb filter. This crosstalk represents a very small difference between the distorted chroma signal and the crosstalk information itself and is not detectable on the TV. However, if a comb filter were not used, crosstalk would be noticeable on the screen.

VHS Crosstalk Elimination

The VHS system uses the same basic technique to accomplish color-under chroma recording and to eliminate adjacent channel chroma crosstalk during playback. However, there are major differences in the way the task is performed. The conversion frequencies are different, but the major difference is in the way that the subcarrier phase is modified in the color-under conversion process.

VHS modifies the subcarrier phase on each line of both the A head field and the B head field. The phase is advanced 90 degrees from each previous horizontal line for every line in the A head field and is retarded 90 degrees with reference to each previous horizontal line for every line in the B head field. The phase rotation information is contained in the 4.21 MHz color-under conversion CW signal just as the phase modification information was contained in the 4.27 MHz signal in the Beta format.

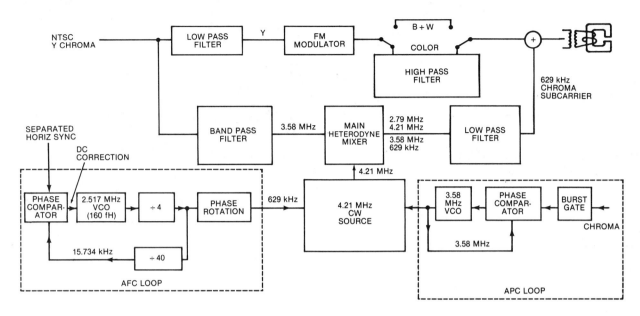

Fig. 5.63. VHS chroma record circuitry.

The VHS phase rotation process is more complex than for Beta units.

Fig. 5.63 is a block diagram of the basic VHS color-under record circuitry. The NTSC combined Y and chroma signals enter a low pass filter as shown in the upper left of the diagram. The low pass filter allows only Y information to pass so it can be frequency modulated and sent to the video head for recording. The combined Y and chroma signal is also passed to a band pass filter where only the higher chroma frequencies are passed. The 3.58 MHz chroma subcarrier signal is heterodyned with a 4.21 MHz CW signal and converted to the 629 kHz difference signal. This color-under subcarrier information is recorded directly onto the tape using the FM Y signal as bias.

The 4.21 MHz CW signal is created from the sum of two signals—629 kHz from the AFC loop and 3.58 MHz from the APC loop. A 2.517 MHz voltage controlled oscillator produces a signal that is 160 times the horizontal sync (160 fH). It is divided by four to yield 629 kHz. The 629 kHz output of the divide-by-four circuit splits to provide inputs to a phase rotation circuit generating the 629 kHz output for the 4.21 MHz CW source and to a divide-by-forty circuit that generates a 15.734 kHz input to a phase comparator. The phase comparator matches the separated horizontal sync input with the 15.734 kHz input to generate a DC correction voltage for phase locking the 2.517 MHz VCO with the separated horizontal sync pulses from the incoming video.

The other input to the 4.21 MHz subconverter is 3.58 MHz from the Automatic Phase Control (APC) loop. A 3.58 MHz VCO produces this CW signal. The input to the APC circuit is a sample of incoming record chroma signals. The APC identifies the incoming video's chroma burst, compares its phase with the phase of the 3.58 MHz VCO and makes any corrections needed by causing a DC control voltage change at the 3.58 MHz VCO frequency and phase-locking the color conversion. The color 4.21 MHz CW conversion signal is thus frequency- and phase-locked to the incoming video signal.

Inside the AFC loop is the phase rotation circuitry shown in Fig. 5.64. Incoming separated horizontal sync pulses enter the AFC phase comparator and are matched with a 15.734 kHz input to generate a DC control voltage for the 2.517 MHz (160 fH) VCO. The 160 fH VCO output is passed to a ring counter where it is divided by four (4 fH) and each output is shifted 90 degrees. The outputs described as signals Phase 1, Phase 2, Phase 3, and Phase 4 are phase shifted 90 degrees apart from each other. The Phase 2 output is divided by forty to produce a 15.734 kHz (the horizontal sync rate) input to the AFC phase comparator to lock the 2.517 MHz VCO on frequency.

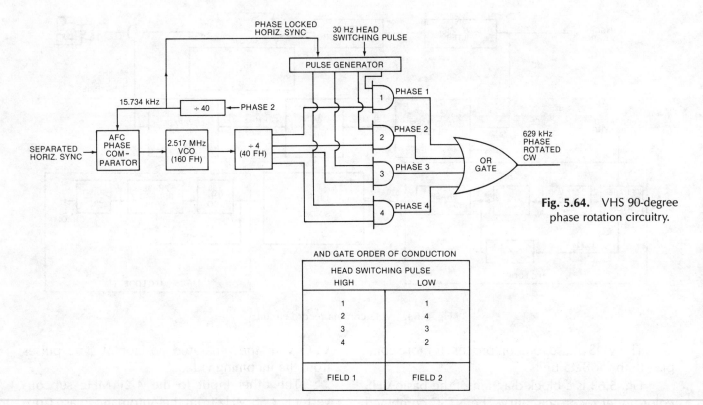

Fig. 5.64. VHS 90-degree phase rotation circuitry.

The outputs from the divide-by-4 ring counter are fed into some two-input AND gates that produce an output only when each input is a logic level 1 (high). Thus the 629 kHz phase-shifted signal does not pass through one of the AND gates until its other output is also high. The second input to the AND gates comes from a pulse generator enabling circuit. The AND gate Pulse Generator selects which of the four output lines to enable according to the phase-locked horizontal sync and the 30 Hz head switching pulse applied. Since each of the AND gates at the ring counter output has a phase-shifted 629 kHz signal on one input at all times, all that is needed to allow that signal to pass through is a logic high from the appropriate gate in the AND gate Pulse Generator.

The enabling or firing order from the AND gate Pulse Generator is shown in the lower right of Fig. 5.64. The 30 Hz head switching pulse is used to alter the order of firing. The horizontal sync input causes the AND gate Pulse Generator to produce an output at the horizontal scan rate.

The outputs from the four AND gates are fed into a four-input OR gate that passes any signal present at any input. It doesn't care if it has a signal present at one input or all of them; it will pass one or a composite through to its output. While head A is recording, the first horizontal line AND gate 1 is turned on by signals at both of its inputs. It passes a zero degree phase-shifted signal to the input of the OR gate. All of the other AND gates are disabled. The OR gate passes the signal from AND gate 1 to the 4.21 MHz subconverter and the 3.58 MHz chroma signal is phase-altered accordingly. The next horizontal line comes along and AND gate 2 is enabled. At the next firing of the Pulse Generator, AND gate 2 outputs a +90 degree phase-shifted signal through the four-input OR gate to the CW source which causes the 3.58 MHz sub-carrier phase to alter +90 degrees as it is converted to the color under frequency. Phase sequencing occurs for each line in that field.

During head B record, the same process takes place, but now the AND gates are enabled in a different order because the 30 Hz head switching pulse has changed polarity. This causes a different sequence of commands from the AND gate Pulse Generator and the firing order is such that the phase is retarded 90 degrees rather than advanced.

In playback a similar procedure is followed except the incoming color-under chroma has time base instabilities in it which must be corrected.

Magnetic Recording Theory   133

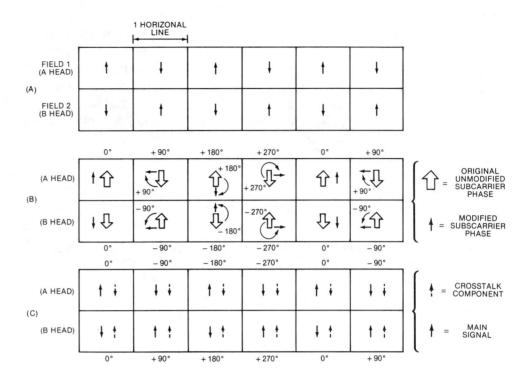

Fig. 5.65. VHS chroma phase relationship before and after record and playback.

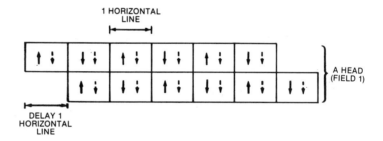

Fig. 5.66. VHS main subcarrier and crosstalk phase relationship after signal is delayed 1 line.

This is done the same way as it was for the Beta units. The AFC's incoming horizontal sync is now coming from the demodulated FM which includes time base errors. Another difference is that the phase relationships are restored to their original state by the Rotational Phase Converter. Head A was recorded with the chroma phase advanced by 90 degrees. To restore the chroma to its original phase during playback, the A head information must be retarded by 90 degrees. The B head phase was retarded by 90 degrees in record so it is advanced +90 degrees in playback. Crosstalk from the adjacent channels is also phase-rotated with the main chroma signal. The up-converted 3.58 MHz chroma containing the crosstalk is applied to a comb filter where the crosstalk information is reduced to near zero just as it is in Beta machines.

Fig. 5.65A shows the chroma phase relationship for the two vertical fields; Fig. 5.65B shows the rotated phase relationship after the chroma has been down converted, and Fig. 5.65C shows the restored phase after it has been up converted during playback. Section C also shows adjacent track crosstalk. Notice how it has been phase-rotated.

A comparison between Fig. 5.66 and Fig. 5.62B (Beta crosstalk phase relationships) shows that the signals produced by the two methods are similar. The comb filter cancels the crosstalk portion of the signal as described for Beta machines.

After examining the diagrams of the modified chroma phase relationships for both formats, one might wonder why the VHS crosstalk is 180 degrees out of phase with the other field when it is not for the Beta. (Compare Fig. 5.61C with Fig. 5.65C.) This is caused by the simplification of the diagrams. They do not reflect the exact NTSC (National Television Systems Committee) interlace format but are simplified to help you understand basic concepts. The point is that the comb filter cancels signals that have the same phase relationships within any given horizontal line of information when that line is delayed and added to the same nondelayed line of information. Cancellation occurs regardless of the signal polarity as long as they are in phase. Another interesting feature of the comb filter is that it has the ability to add or subtract signals at its output by simply reversing the input or output transducer leads. It is therefore possible to make out-of-phase signals cancel and in-phase signals add (just the opposite occurs in Beta).

These circuits are extremely complex and proper alignment is essential to maintain correct operation and reproduce crisp, clear color video. Troubleshooting the color circuits can be challenging but also fun if you know what to look for. Since most of the circuit design and signal processing is now contained inside integrated circuits (ICs), troubleshooting is much simpler. It is also worth noting that ICs are quite reliable so most VCR problems are mechanical or due to operator error.

This chapter has covered audio and video signal recording on magnetic tape and the means by which the signal is recovered and made primarily noise and crosstalk free. Much more of the operating electronics needs to be covered including more detail on the color circuits. The next chapter will describe how the VCR operates at the circuit level.

CHAPTER 6

VCR Operating Theory

In the last chapter, you discovered how magnetic principles are used to record and recover information that is stored on a special oxide-coated plastic length of tape. Chapter 6 describes the complex circuitry that surrounds the magnetics and enables audio and video to be sensed and converted into analog and digital signals to produce a beautiful color program on a TV or monitor screen. This portion of the book begins with a look at color signal processing.

VCR color signal processing involves more than frequency conversion. The machine must be able to deal with color levels that vary from one program to the next, and color intensity or saturation levels that vary as different channels are selected on the tuner. During playback, each video head may respond differently to the color signals recorded on the tape, producing different color levels between the A and B heads. During black and white recording, the VCR must deactivate the color processing circuitry to eliminate color noise. If the color processing circuitry were left on while recording a black and white program, color noise would appear as color beat patterns or as random color speckles across the screen as the unit tries to produce a color picture from a black and white signal stream. Many VCRs allow recording of the full lower side band for the Y FM signal during a black and white program. This is another reason why the VCR must be able to decide whether the program is black and white or color and respond accordingly.

AUTOMATIC COLOR CONTROL (ACC)

In the Beta machines, the burst portion of the chroma signal is important for chroma level control. Automatic color control (ACC) occurs during recording at the input to the main heterodyne mixer circuit. Figure 6.1 is a block diagram of the early Beta record color processing circuitry. ACC acts on the original 3.58 MHz chroma signal. The ACC variable gain amplifier can adjust its gain very rapidly to stabilize the varying levels of incoming chroma signal. The control signal for this variable gain amplifier is derived by sampling the burst signal because it represents the most stable part of the color signal. Chroma levels within the picture viewing area vary in saturation, but the burst signal is relatively constant. A burst from the incoming

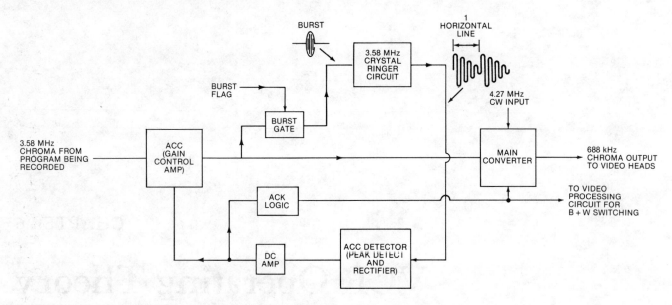

Fig. 6.1. Early Beta ACC and ACK record processing.

chroma signal is identified by a pulse known as a *burst flag* or *burst gate*. Burst flag is the horizontal sync signal which has been separated from the video signal and delayed slightly for positioning.

The burst pulse is gated and applied to a crystal ringer circuit which oscillates at a frequency determined by the crystal when it has excitation pulses of the same frequency applied. One-and-a-half cycles of burst are applied to this crystal ringing circuit to cause excitation and oscillation until the next burst pulses arrive. The amplitude of the ringing circuit output oscillation varies according to the level of the incoming burst signal. A peak detection circuit samples the output oscillation from the ringing circuit, and converts the waveform into a DC control voltage that is applied to the ACC.

When the system is first energized, (or if no input chroma is present) the ACC operates at full gain and adjusts accordingly when color signals arrive at the input. During record, the circuitry assumes that the incoming chroma level is relatively stable, so the ACC response is not as fast as it is during playback. In earlier models, the 3.58 MHz ringing circuit uses a separate crystal oscillator to cause ringing. Recent technology has provided methods for ACC control using fewer components, and ACC is now performed almost entirely in a single IC.

During playback, the ACC process is more complicated as shown in Fig. 6.2. Methods for playback chroma level sensing and stabilization are similar to that for record, but the incoming color has greater potential for changes in level. Level differences are caused by different playback amplitudes from the video heads and chroma level variations from recording and playback. Playback chroma level can vary from one line to the next as well as from one field to the next. If the playback color level varies substantially between the A and B heads, the TV screen has a 30 Hz color flicker. Color flicker can vary from slight color differences between frames to good color for one frame and no color for the next. Beta playback ACC counters the 30 Hz flicker with two separate ACC control voltage storage networks. One storage filter network is for the A head, the second is for the B head. As the A head crosses the tape, the appropriate ACC control voltage is stored in the network applied to the ACC circuit. When the B head crosses the tape the second storage network is switched in and becomes active. ACC compensates for differences in chroma playback between the two separate heads.

In record, the ACC circuit passes the level controlled signal to the main heterodyne mixer circuit where it is converted a 688 kHz color-under signal. The 688 kHz color-under signal is then applied to a burst emphasis circuit to increase the amplitude of burst relative to the remainder of the

VCR Operating Theory   **137**

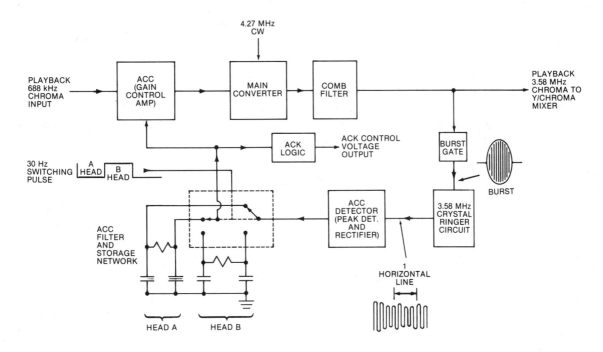

Fig. 6.2. Early Beta playback ACC with separate ACC storage and filter networks.

1.5 dB (decibel) chroma signal in the Beta 2 record mode. The burst in Beta 3 is boosted by 6 dB.

An increase of 6 dB is equal to a doubling of the voltage. A 6 dB increase in a one-volt peak-to-peak signal means that the original signal now equals 2 volts peak-to-peak. A 3 dB gain equates to one-and-a-half times increase in the normal signal.

Burst emphasis increases the burst signal during record. During playback, burst affects ACC and the elimination of time base errors in the color-up process. The burst signal is emphasized in record to provide better signal-to-noise-ratio during playback. A burst signal of increased amplitude enables better ACC and control of time base instabilities. Boosted burst is also less sensitive to signal deterioration while recording and then playing back the video signal. During Beta III playback, the emphasized burst signal is de-emphasized by 4.5 dB after up-conversion to 3.58 MHz. Beta II playback burst is not de-emphasized.

AUTOMATIC COLOR KILLER (ACK)

An earlier section stated that the VCR chroma circuits must be de-energized during the record and playback process. During record and playback of a black and white video signal, an automatic color killer (ACK) is used to eliminate output from the chroma processing circuitry and to prevent the VCR from attempting to produce a color program from black and white signals. A failure in this circuitry could produce color snow or other forms of random color noise on the screen during playback.

Early Beta units used ACK to generate a control voltage to turn off the main heterodyne mixer circuit. The color killer circuit monitors a DC voltage on the input to the ACK section. ACK has a threshold level that is adjusted during alignment. Once that threshold is reached, the ACC control voltage causes the main heterodyne mixer circuit to turn off. If the ACK is activated during record, it produces a color killer voltage which permits use of the full lower FM sideband and increases picture resolution.

The control voltage for the ACC and ACK signals is derived from the peak detect circuit. Since the peak detect circuit is driven by a 3.58 MHz crystal ringing circuit and only oscillates when there is burst present from the incoming video signal, ACK can be directly associated with the presence of burst on the record signal. Recent ACK technology squelches the chroma output signal rather than disabling the main heterodyne mixer circuit. The squelching process affects the chroma signal from the main heterodyne mixer circuit. This prevents random color noise from being mixed with the Y FM signal and then being recorded.

In playback, ACC monitors the burst signal after the crosstalk component has been removed from the up-converted chroma signal. Chroma burst is sampled using a burst gate just prior to the YC mixer circuit which produces the final machine output for use by the TV. Burst excites the crystal ringer so the amplitude of the oscillations are peak-detected and rectified to a DC control voltage for the ACC circuitry.

As mentioned earlier, ACC controls the amplitude of the incoming 688 kHz chroma signal and varies its amplitude to provide a stabilized chroma level for the main mixer. Two filter and storage networks are provided in the ACC DC control voltage line. One network filters and stores the appropriate DC voltage for head A while the second network filters and stores the voltage for head B. An electronic switch is toggled by the incoming video head switching pulse to connect the appropriate filter and storage network at the correct moment.

This circuit produces differences in the chroma playback signal between the two video heads. Playback ACC doesn't just compensate for chroma differences on a line-by-line basis; it also compensates for these differences on a field-by-field basis. The same crystal ringing circuit is used for both playback and record.

THE BID CIRCUIT

During playback there is a possibility that the 4.27 MHz phase inverter circuit can lose sync and get out of phase. The color circuitry ensures that the APC circuit is not operating 180 degrees out of phase. If APC were to operate 180 degrees out of phase from the original recorded signal, severe color differences would occur between the record and playback hues. Excessive color flicker could also be present on the screen. A burst identification circuit (BID) is used to prevent these potential problems as shown in Fig. 6.3.

The 4.27 MHz phase inverter circuit is driven by phase inverter selection logic in BOTH record and playback. The phase inverter selection logic is comprised of a *flip-flop* and two OR gates. (The operation of an OR gate has previously been described.) A flip-flop is a digital logic circuit whose output voltage changes when an input signal (such as a clock pulse) is applied. It maintains the output voltage until another input clock signal is applied causing it to change back to its previous state and output.

The input to this flip-flop comes from an OR gate with two inputs. One of these inputs is the horizontal synchronization pulse. The second input is the BID signal. The output of the flip-flop is passed to a second two-input OR gate. The other input of this gate is a 30 Hz video head switching pulse. The output of the second OR gate drives the 4.27 MHz inverter selector circuitry.

During playback of video head A, the video head switching pulse is low and has no effect on the OR gate output. Horizontal synchronization pulses are passed to OR gate 1 and to the flip-flop. When the first horizontal pulse occurs, the flip-flop is toggled high and remains high until the next horizontal synchronization pulse arrives to cause the flip-flop to toggle low. The output of the flip-flop is felt at one input to OR gate 2. When the other input (video head switching pulse) to OR gate 2 is low, the flip-flop state changes to match the output of OR gate 2.

The flip-flop output alternates at the horizontal rate and is passed to the phase inverter selection circuit. The horizontal synchronization pulse input causes this circuit to toggle between the inverting and the noninverting input. When the video head switching pulse goes high, head B is in contact with the tape. The logic high video head switching pulse is applied to the input of OR gate 2. Because this input is high, the output of OR gate 2 remains high regardless of the flip-flop state. The output of OR gate 2 forces the phase inverter selection circuit to select the noninverting input during the time that the video head switching pulse is high.

During playback, it's important that the modified subcarrier phase be converted back to its exact prerecorded phase. During the time that head A is recording, the subcarrier phase is modified so each adjacent horizontal line has a subcarrier of the same phase. This subcarrier phase must be reconstructed exactly as it was during playback for the TV to properly recover color. If the phase in-

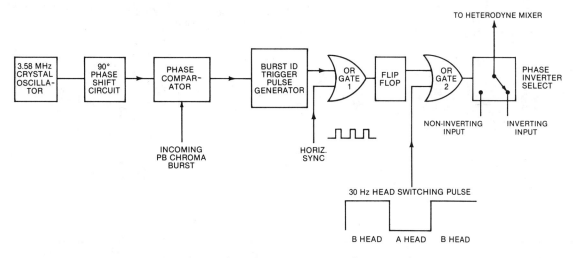

Fig. 6.3. Early Beta Burst ID (BID) pulse generator.

verter selection circuit were operating 180 degrees out of phase, proper color recovery would be impossible. This circuit can invert the wrong horizontal line subcarrier if the flip-flop operates out of phase. The flip-flop changes state every time a horizontal synchronization pulse appears at its input. Missed reading of horizontal lines due to tape damage or missing oxide particles on the tape surface is called *drop out*. If drop out occurs and a horizontal synchronization pulse were not read off the tape, the flip-flop would remain in its current state until the next horizontal line appeared causing the phase inverter selection circuit to invert the wrong line.

In Fig. 6.3, the BID pulse is the second input to OR gate 1. The burst ID circuit resets the flip-flop any time incorrect chroma phase is detected by applying a positive going pulse to the OR gate causing the flip-flop to toggle.

There are three basic inputs to the BID circuit. These are: burst gate, playback 3.58 MHz up-converted, chroma, and reference 3.58 MHz from the APC crystal oscillator. The BID compares the phase of the playback chroma burst to that of the stable crystal oscillator in the APC circuit. The local 3.58 MHz oscillation is phase shifted 90 degrees and then applied to the phase comparator to position the playback chroma burst in the same relative phase with the local 3.58 MHz oscillation. Whenever a phase inversion between the 3.58 MHz local oscillation and the burst signal is detected, a positive going pulse is passed from BID OR gate 1. This pulse toggles the flip-flop to the correct state. The BID outputs a signal only when a 180-degree phase error is detected. Minor phase errors are corrected by the playback AFC and APC circuits.

Technological advances have simplified the chroma process circuitry in the Beta systems. Fig. 6.4 shows a comparison between the current and previous Beta color playback processing circuits. The ACC, BID and ACK circuits, and the method for development of the 4.27 MHz frequency conversion signal are slightly changed from the original designs. APC has been simplified for both record and playback.

The figure shows color signal processing for past and current Beta machines. An obvious difference is the change in the AFC loop VCO and the APC VCO frequencies. One of the most significant changes is the simplification of playback chroma processing. The first Beta VCRs required both AFC and APC for color-up conversion. Current technology does not use AFC to eliminate chroma playback time base errors. These are eliminated by the use of APC and an advanced BID circuit. In current machines, BID is called automatic phase control ID (APC-ID).

There have also been improvements in the use of APC in the color-under conversion process during record. Digital processing techniques and higher VCO frequencies to derive the 4.27 MHz color conversion signal enable simplified APC during record, while maintaining high accuracy. The important point to remember is that the end result is the same. The color signal being processed must

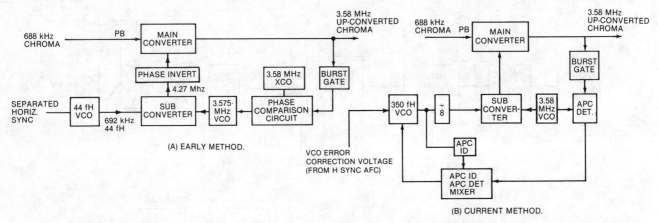

Fig. 6.4. Comparison of Beta playback chroma processsing techniques.

maintain proper signal level as controlled by ACC. It must be identified as black and white or color, and it must be identified by ACK circuit, be recorded on the tape at 688 kHz and have every horizontal line within the A field to maintain the same chroma subcarrier phase relationship. During playback, time base errors must be removed, and the original subcarrier phase relationships must be restored. Procedures may have been changed but the basic color recording process remains the same. Also, many of these tasks are now accomplished using integrated circuits instead of transistors, resistors and other discrete components.

CURRENT CHROMA SIGNAL PROCESSING DURING RECORD

The block diagram in Fig. 6.5 describes current chroma signal processing for record. It complements Fig. 5.59 from Chapter 5. The heart of the AFC record circuit is a 350 fH VCO (5.5 MHz) signal. The output of the 350 fH VCO is passed through a divide-by-eight circuit to produce 688 kHz (43.75 fH). The VCO is controlled by the AFC feedback loop. Horizontal sync is separated from the incoming record video and applied to an AFC detect circuit that monitors the 350 fH VCO after its frequency has been divided by 70. The AFC circuit compares incoming horizontal sync with a 5 fH sample from the VCO. Frequency errors in the VCO are detected producing a control voltage from the AFC detect circuit that adjusts the 350 fH VCO to achieve the correct frequency output. The 688 kHz signal from the divide-by-eight circuit is fed directly into the heterodyne mixer circuit subconverter. In earlier machines, this signal was exactly 44 fH (692 kHz). The second input to the heterodyne mixer circuit subconverter is a stable crystal controlled 3.58 MHz signal. Earlier machines use a 3.575 MHz signal. 688 kHz is added to 3.58 MHz to produce 4.27 MHz. This signal is phase inverted every other line during head A recording and applied to the main heterodyne mixer.

The 3.58 MHz variable crystal oscillator (VXO) phase can be varied by a phase detect circuit. In record, APC detect is used to sample the burst of the incoming 3.58 MHz chroma signal. The APC detect circuit compares the output of the 3.58 MHz VXO with the incoming chroma burst and generates a DC control voltage to vary the phase of the VXO. This phase controlled 3.58 MHz signal is then passed to the heterodyne mixer circuit subconverter. The 4.27 MHz chroma conversion signal is now frequency and phase locked with the incoming video signal.

The APC detect circuit also sends a control voltage to the ACC detect circuit. ACC detect compares the APC control voltage with a sample of the burst gate information and applies a control voltage to the ACC amplifier. The ACC circuit then adjusts for stable chroma input levels as directed by the APC detect circuit. The ACC detect circuit output is monitored by the auto color killer circuit. ACK monitors the ACC detect circuit and compares its output voltage to a threshold level established during alignment. Once that threshold level has been reached, ACK generates a DC control voltage for the Y chroma mix circuit preventing er-

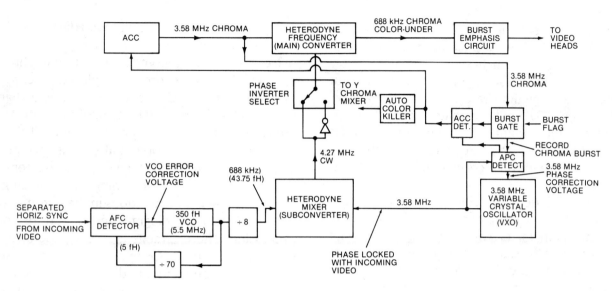

Fig. 6.5. Current Beta chroma record.

roneous color noise from entering a black and white program.

During chroma signal playback, time base instabilities must be removed, and the 688 kHz color-under signal must be converted to 3.58 MHz. Fig. 6.6 is a block diagram of this process. Playback 688 kHz chroma enters the ACC after having been separated from the Y FM signal. The color-under signal is up-converted to 3.58 MHz in the main converter. The 4.27 MHz color-up conversion signal comes from the subconverter. This signal is both frequency and phase corrected.

Incoming horizontal sync is not used in current Beta AFC as in earlier Beta machines. The 350 fH VCO is frequency controlled by the APC-ID. The 350 fH VCO output is divided by eight and a

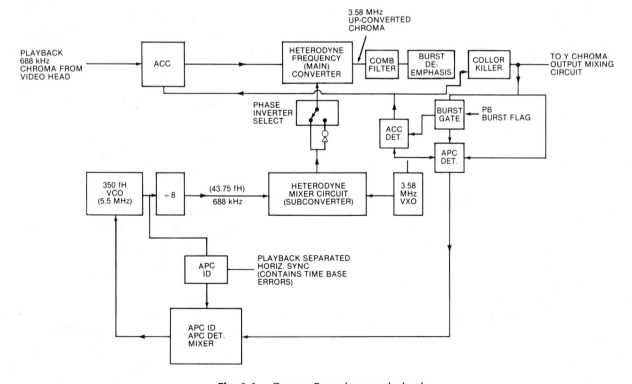

Fig. 6.6. Current Beta chroma playback.

sample of the resultant 688 KHz signal is sent to the APC-ID circuit. The APC-ID monitors playback horizontal sync which has been separated from playback video. This incoming horizontal sync contains the large time base errors in the playback signal. The APC-ID only generates a correction signal when large horizontal frequency errors are detected. Otherwise, this circuit relies on the APC detect to make chroma corrections and eliminate time base errors.

The APC detect circuit compares the burst from the 3.58 MHz up-converted chroma signal from the main mixer with the stable 3.58 MHz VXO. The chroma signal is sampled in playback after it has been up-converted and passed through the comb filter. Chroma time base errors are detected when they are compared with the stable 3.58 MHz VXO, and a control signal is applied to the APC-ID and APC detect mixer circuits.

Two points of interest should be noted. First, the 3.58 MHz VXO functions as a stable crystal oscillator in playback. It is not phase altered by the APC detect circuit as it was during record. Secondly, playback phase correction is performed inside the 350 fH VCO and controlled by the APC detect circuit. APC detect can only make phase corrections for the elimination of time base errors by + or − 1 fH (15.734 kHz). If the 350 fH VCO frequency error is greater than 1 fH, detection by the APC-ID circuit causes it to take control of the 350 fH VCO and adjust the 350 fH VCO back to the correct frequency. APC-ID then returns control to the APC detect circuit.

The APC-ID circuit is a digital counter that operates in the following manner. An incoming heterodyne subconverter frequency error of 1 fH causes a frequency error of 8 fH in the VCO. This large frequency discrepancy is detected by APC-ID when it compares playback separated horizontal sync with the VCO frequency. The APC-ID circuit functions like the AFC circuit in former circuits. The difference is that this AFC circuit is digital.

Some current Beta machines use a 175 fH VCO and a divide-by-four circuit instead of 350 fH. APC detect and APC-ID circuits can be found in these machines.

VHS Chroma

VHS chroma circuits have remained relatively unchanged over the years although digital technology, as outlined in Fig. 6.7 and Fig. 6.8, has replaced the analog components in the chroma circuits. During record and playback chroma processing, the AFC circuit operates as a digital phase-locked-loop (D-PLL). One of the features of the D-PLL is that it can act as a buffer oscillator in the event playback horizontal sync is missing. It can be considered an artificial sync pulse generator that functions when incoming sync is missing or incorrect. The D-PLL functions in the chroma AFC circuits. The block diagram shows current color processing. These circuits are nearly identical to original VHS circuits except that color conversion processing frequencies are digitally generated and the block diagrams now include the ACK and ACC circuits.

VHS ACK and ACC signal processing is similar to that described for Beta. One major difference concerns burst boost. VHS machines amplify the burst signal 6 dB in the SP and SLP (EP) record modes. Burst is not amplified in LP. It was discovered that the color signal-to-noise ratio increased when burst was amplified (boosted) during selected recording speeds.

In LP and SLP, video track overlap occurs. Horizontal sync and burst are aligned on the two adjacent tracks. The adjacent channel horizontal sync and burst signals beat together to produce a form of crosstalk. In the SP and SLP speeds these beats occur where horizontal sync and burst appear in the video signal. This means they are not seen.

There is no sync alignment in LP. The adjacent channel horizontal sync and burst beat pattern are no longer contained in a screen area out of view. Instead, the beat occurs within the picture information itself. To keep this LP beat pattern as low as possible, the burst level is not boosted during LP record. This results in a consistent beat screen intensity and is considered least objectionable.

If the beat pattern intensity were to vary, attention would be drawn to the interference on the screen. The chroma signal-to-noise improvement produced in the color processing circuitry by boosting the burst is insignificant in LP when the sync beat patterns are also increased. Since the sync beat patterns occur off the screen in SP and

VCR Operating Theory   143

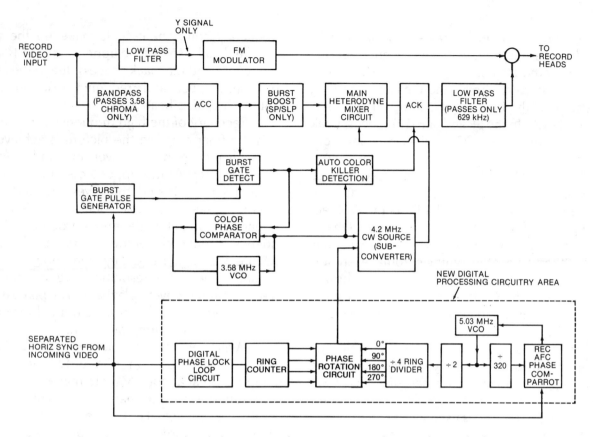

Fig. 6.7. VHS digital chroma record processing is similar to earlier methods.

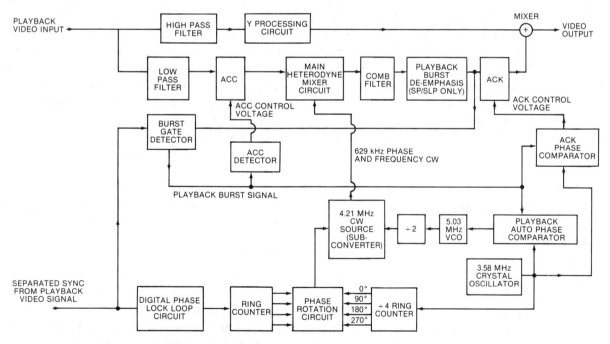

Fig. 6.8. Current VHS chroma playback is similar to earlier methods except digital processing has replaced analog processing.

SLP, the burst can be increased for these two speeds increasing the color signal-to-noise ratio.

The method to achieve ACC in VHS machines also varies between the LP and the two remaining speeds. In LP, the burst level is sampled and used to control ACC. In the SP and SLP (EP) speeds, peak detection is used. LP ACC requires burst level detection to assure that the burst level is held constant to reduce the beat pattern from sync pulse misalignment. In the SP and SLP modes, ACC detects peak chroma levels and adjusts the ACC accordingly. Burst is a part of the chroma signal that is peak detected during SP and SLP. Burst often acts as the highest level of incoming chroma, even during SP and SLP record modes. Because chroma levels are very rarely saturated in the broadcast signal, burst is often a significant part of the peak level detect for ACC in these SP and SLP modes. The purpose for these two methods for accomplishing ACC in the various speeds is to provide maximum signal-to-noise ratio in color processing. Peak level detection provides the best signal-to-noise ratio for chroma processing and is used whenever possible.

The introduction of audio FM recording has caused the necessity for more traps to isolate the chroma signal in playback. Audio FM recording and playback will be described later. It's important to understand that audio FM recording uses FM frequencies near the color-under signal. Additional filtering is required to separate the playback color-under signal from the audio FM. These filter circuits are required in both formats.

In playback, the amplified burst signal is restored to normal after the signal has been up-converted.

FM Luminance

VCRs use frequency modulation to record the luminance portion of the signal as described in Chapter 5. In VHS units, the modulating frequency varies between 3.4 MHz (corresponding to "sync tip"), to 4.4 MHz (corresponding to 100 percent white). Standard Beta units use an FM deviation of l.3 MHz, ranging from 3.5 MHz for sync tip to 4.8 MHz for peak white. Most of the RF (radio frequency) information contains the picture detail and falls within this deviation range. Much of the luminance signal fine detail is contained within the sidebands and must be retained to reclaim the high frequency components that make up the video picture detail. Most of the upper sideband is lost in the record and playback process due to the effects of writing speed and frequency response losses within the video head itself.

Recovery of the high frequencies which contain the finer detail in the picture is achieved by carefully recovering the lower sideband. Since the ability to recover the upper side-band of the recorded signal is limited, the lower sideband is used to compensate for absence of most of the high frequency signal. Before the prerecorded signal is frequency modulated, compensation is performed to restore the high frequency components and to minimize the loss of resolution due to limited upper-sideband recovery playback. The playback signal is de-emphasized to restore normal video. De-emphasis occurs during playback.

While recording a color video program, the extreme lower portion of the FM sideband is chopped off at about 1 MHz to make room for the color-under signal. In many VCRs, when a black-and-white-only signal is detected, the full lower FM sideband is recorded on the tape providing more picture detail.

Fig. 6.9 is a block diagram that describes the processing steps for the Y portion of the video signal during record. This signal first encounters a low pass filter that separates the Y and chroma signals. The chroma information consists of the high frequency portion of the video signal. Picture brightness and detail information are contained in the lower frequency portion of the video band. The lower frequency video information is called *luminence* and designated with a Y.

Comb filters are being used today as VCR low pass filters. They provide excellent chroma and Y signal separation while preserving much of the high frequency portion of the Y signal.

Y AGC

Separated Y must be level controlled before it can be modulated. Video levels vary substantially between recorded programs. To provide proper signal reproduction, the FM modulator must have a video signal applied which varies within limitations regardless of the level of signal coming into the VCR. *AGC* (Automatic Gain Control) ensures that the video signal in the modulator is within re-

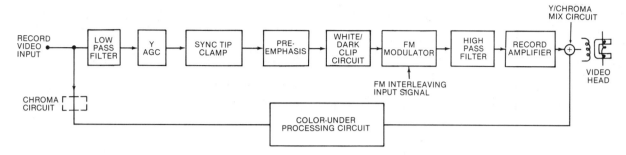

Fig. 6.9. The Y FM modulator.

quired limitations. Incoming video must be stabilized to 1 volt peak-to-peak between sync tip and 100 percent peak white.

The Y AGC circuitry performs complicated tasks by ensuring that the incoming video signal is within the specific limitations required by the modulator. However, it cannot react to normal variations in the reproduction of scene brightness. If the AGC were simply to detect and limit peak video levels, normal peak brightness would be reduced while average dark scenes would have increased brightness levels. This would produce normal scenes with brightness levels much darker than they should be while dark scenes would have unusually high levels of brightness.

AGC is designed to correct for errors in overall amplitude. It does not correct for normal brightness variations. To accomplish its task, AGC monitors and responds to variations in the horizontal synchronization pulse. In a one volt peak-to-peak standard video signal, horizontal sync is fixed at a constant 0.3 V amplitude. This leaves the remaining 0.7 V for picture brightness levels. Picture brightness may vary with scene content but the sync amplitude levels must remain constant. Sync amplitude detection is a stable way to monitor video level errors in the incoming video signal.

Sync detect AGC is achieved by separating the incoming video horizontal sync, inverting, amplifying and delaying the sync pulse, and reinserting the pulse in the horizontal blanking interval. This inverted and amplified horizontal sync pulse is applied during the horizontal blanking interval so it won't interfere with video information presentation. This inverted pulse is only used in the video AGC circuit for level detection and is not recorded on the tape. The amount of amplification that this inverted and delayed horizontal sync pulse re-

ceives is enough to produce 1 V peak-to-peak between sync tip level and the positive going peak of the inverted sync pulse.

Fig. 6.10 shows the relationship between the incoming video signal containing both Y and chroma information, the Y-only signal after color has been filtered out, the separated horizontal sync, the horizontally delayed, amplified and inverted horizontal sync, and the combined signal applied to the AGC detector. The AGC detector monitors the relationship between the horizontal sync tip level and the internally manufactured 100 percent peak white reference level. It also adjusts the AGC amplifier to maintain a 1 V peak-to-peak detection level. As the video signal levels from the tuner vary between stations, or the incoming recorded video signals differ between video cameras, the AGC makes the appropriate corrections to maintain the 1 V peak-to-peak reference.

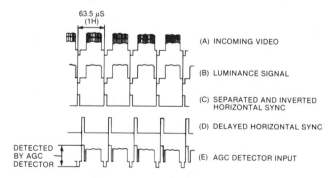

Fig. 6.10. AGC circuit waveforms.

The AGC detector is a peak detection circuit that is not influenced by normal variations in video scene brightness. If the video level content exceeds the normal 0.7 V peak-to-peak level, the AGC detection circuit recognizes that the video level has exceeded the reference 1 V peak-to-peak

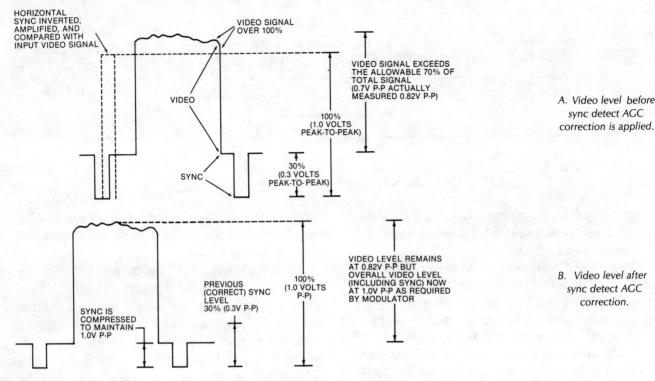

Fig. 6.11. AGC compresses sync.

signal internally generated in the machine. As peak level is detected, the AGC amplifier begins to compress the sync level signal applied to the modulator. When the video portion of the input signal is more than the 100 percent peak white reference level, the AGC circuit acts to keep the video component at a constant amplitude. To do this, the AGC must compress the sync level slightly and add this compressed pulse level to the 100 percent peak white reference level to keep the total video information within the 1 V peak-to-peak signal required by the modulator. Fig. 6.11A describes the video level before sync detect AGC correction is applied. Fig. 6.11B shows the video level after AGC correction.

Sync Tip Clamp

The functional block you saw to the right of Y AGC in Fig. 6.9 is the Sync Tip Clamp circuitry. Once the incoming video signal has been level-corrected by the AGC circuit, the video sync tip is tied to a predetermined DC voltage in the clamping circuit. Using sync tip forces the modulator to the lowest frequency in its deviation. Sync tip must not go below a certain prefixed voltage. If it did, the modulator would cause a deviation below the desired frequency. Clamping is necessary to prevent the sync tip from going too negative. Some circuits use the sync tip clamp in the Y record process circuitry prior to AGC detection.

Pre-emphasis

In the processing circuitry of Fig. 6.9, the next block (Pre-emphasis) is used to increase the high frequency response of the video recorder and to increase the signal-to-noise ratio. Pre-emphasis levels must be carefully selected to maintain suitable signal-to-noise ratio and to minimize the effects of spurious noise caused by too much pre-emphasis. If too much pre-emphasis is applied, crosstalk from adjacent video tracks is increased. The trick is to emphasize the high frequency portions of the signal just enough to provide good playback frequency response and yet keep the signal-to-noise ratio as high as possible. Fig. 6.12 and 6.13 show an actual pre-emphasized signal and its relationship to the FM deviation frequencies.

Pre-emphasis and its relationship to the video signal is shown in Fig. 6.14. From left to right, notice the horizontal sync pulse that rises to the video reference black level. This is known as the video *set-up* or *pedestal level*. The set-up level corresponds

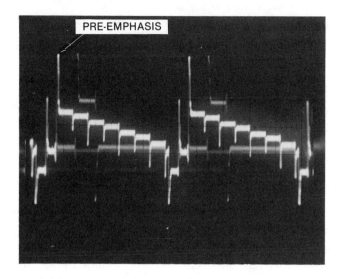

Fig. 6.12. Video waveform with pre-emphasis.

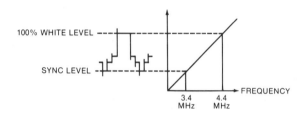

Fig. 6.13. Relationship between pre-emphasis and FM modulation.

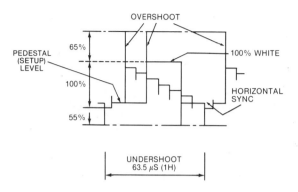

Fig. 6.14. Relationship between the video signal and pre-emphasis.

to pure black as viewed on the TV screen. There is an upward spike at the point where the video level rises from sync tip to the set-up level. This spike is the applied pre-emphasis signal. Notice in the video information portion of the waveform, that each time the signal goes from a dark level to a white level, the pre-emphasis pulse goes in the same direction. It overshoots the level of the original video circuitry. After the signal has reached the 100 percent white level, it begins to gradually step back down to the video set-up level. This waveform is called a *video stair step*. The pre-emphasis circuit causes negative going spikes to occur each time the video signal goes from a white signal to a darker signal.

Pre-emphasis is used to force the modulator into a new modulating frequency at a rapid rate because of the emphasized excursion in the video level. If the proper amount of pre-emphasis were not used, the playback signal would not have good square corners on the video stair step. These square corners represent the high frequency portions of the video signal. Pre-emphasis is important in the reproduction of picture detail because it plays an important part in the recovery of high frequency loss due to the upper and lower FM sideband limitations.

The pre-emphasized signal is then passed to the white and dark clip sections as shown in Fig. 6.9. White and dark clipping is used to control the amount of pre-emphasized overshoot. Pre-emphasizing the signal creates an overshoot in both the white and dark directions. If these excessive overshoots were allowed to enter the modulator, it could be overdriven resulting in a smeared video display where transitions occur between black and white. The loss of detail would be evident when video scenes change levels of brightness. White clip and dark clip adjustments limit the pre-emphasis pulse excursions as shown in Fig. 6.15.

FM MODULATOR

The FM modulator receives the processed signal and converts it into a signal whose frequency deviates in accordance with the incoming video level. The modulator is a VCO with a frequency controlled by the DC level of the incoming video signal. The VCO has a predetermined free running frequency. This means that it has the ability to oscillate at a certain frequency with no incoming signal. Normally, the free running frequency of the modulator is the same frequency as that of the sync tip level. For VHS machines this frequency is 3.4 MHz; the standard Beta free running frequency is 3.5 MHz.

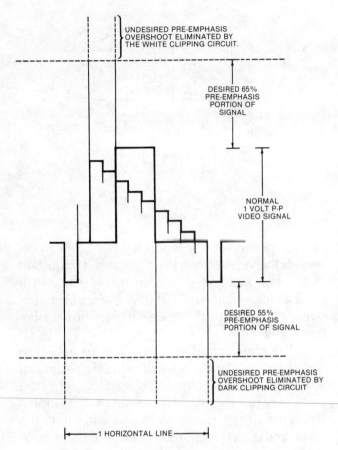

Fig. 6.15. White and dark clipping eliminates excess overshoots.

This frequency is adjusted by eliminating all incoming video signals, placing the unit in record, and electronically adjusting for the required frequency. The DC control voltage entering the modulator is now the sync tip clamp voltage. Incoming horizontal and vertical sync will be clamped to this same reference voltage when they are applied to the modulator.

Symmetry

An important requirement of an FM modulator is that the output waveforms must be symmetrical. Each half cycle of the oscillation must be equal in amplitude and the rising and descending excursions of those waveforms should be the same. This ensures that the lower portion of the FM deviation frequency falls within the original video passband. The demodulated playback video signal contains carrier ripple. This is interference that occurs because the FM deviation frequency and the upper portion of the video passband frequencies are the same.

Carrier ripple appears on the screen as a very fine beat pattern. To eliminate this pattern, the FM demodulator doubles the FM playback frequency prior to demodulation and produces a carrier ripple frequency which is high enough to allow low pass filters to remove this ripple without affecting the frequency response of the playback video signal.

For this double frequency scheme to work, the FM carrier waveform must be symmetrical. If not, the residual carrier energy will cause a fine herringbone pattern on the screen. This form of interference is called *carrier leak*. Adjustments are made during the record modulation process to minimize carrier leak. These adjustments make the FM modulator VCO waveforms as symmetrical as possible.

The Y FM signal is passed to a high pass filter shown in Fig. 6.9. This filter is enabled during the recording of a color program to limit the lower sideband frequencies during record. If a black and white program is being recorded, the filter is bypassed, enabling maximum picture detail on the output of the VCR.

During VHS, LP and SLP speeds, and Beta II and Beta III speeds, the FM modulator frequency is shifted by 7.867 kHz while head B is recording. FM interleaving occurs at these speeds so that the beats from overlapping the tape RF tracks can be made to interleave. The FM frequency difference between the two RF tracks alters the beats produced on the TV screen so they interleave and are not visually detected. The output of the frequency modulator is level-shifted by adding a DC voltage to the incoming video signal during generation of the vertical field corresponding to head B. The video head switching pulse is used to identify the field in the modulator. This alternating frequency is compensated in playback by reversing the process in the demodulator circuit.

FM Record Amplifier

The frequency modulated signal in the record amplifier circuit shown in Fig. 6.9 is boosted in a record amplifier so it can magnetically saturate the video tape.

Another function of the record amplifier is to

provide additional record current during insert edit modes. Insert edit places a new recording in an already existing recorded program. The new video is inserted into the previously recorded signal without disturbing other information on the tape, such as audio.

The full track erase head cannot be turned on during an insert edit or the previously recorded audio and servo control track pulses would also be erased. Since the full track erase head is de-energized, increased Y FM current is applied to the video heads to overcome previously stored video and thus help reduce the color rainbow screen effect. Recall that the video heads do not completely erase the previous video information during record so a full track erase head is used during normal record. Once an insert edit has been performed, a second insert edit cannot successfully be performed on the same section because extra record current applied during the operation will not properly overwrite the previous video signal. It was also recorded with extra record current.

Extra record current is applied under system control of the record amplifier. The extra record current doesn't eliminate all of the rainbow pattern, but it does remove enough so the problem is negligible. Not all VCRs have this feature.

Some frequency equalization is provided in the record amplifier by slightly boosting the low end frequency portion of the signal. This equalization is provided to compensate for the frequency response of the video head during playback as described in Chapter 5.

Record current equalization adjustments are performed in the record amplifier using an *RF sweep generator*. The RF sweep generator is a test instrument which rapidly sweeps through the frequency ranges selected. For VCRs, the sweep generator is required to sweep through the FM deviation frequency range. The sweep generator also provides identification markers at frequencies selected by the technician to enable accurate equalization alignments.

Typically, alignment instructions specify equal frequency response across the FM deviation range. Very few VCRs today provide adjustments for record current equalization. These are factory adjustments only.

After the frequency modulated signal has been voltage and current amplified, the color-under chroma signal is mixed with the FM. The amplitude varying color-under signal requires bias to be recorded onto the tape. The Y FM signal provides this bias.

Rotary Transformer

In order for Y and chroma information to be recorded, the composite signal must pass through the high speed rotating video heads. This signal is applied to the video heads through a rotary transformer comprised of two sections. One section is stationary and is attached to the lower video head drum assembly. The second portion is connected directly to the rotating video head drum. The rotating transformer is a ferrite core with small copper wires wound in such a way that the stationary part of the transformer is able to conduct energy to the rotating portion. Fig. 6.16A and Fig. 6.16B show a cut-away of a rotary transformer and a disassem-

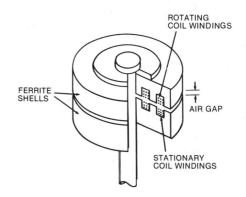

Fig. 6.16A. Cutaway showing rotary transformer.

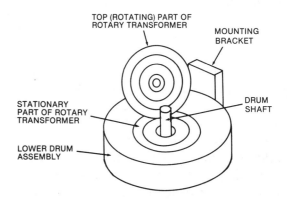

Fig. 6.16B. Cutaway showing rotary transformer but with the rotating portion of the rotary transformer lifted off the drum shaft.

bled video head drum assembly with the rotary transformer exposed. The rotary transformer is not a serviceable item. Failure of the rotary transformer causes replacement of the entire drum assembly. Mechanical precision is required for proper placement of the upper and lower portion of the rotary transformer to provide maximum energy transfer during high speed rotation.

The rotating video head drum assembly is connected to the upper portion of the rotary transformer with very tiny wires. When the video head assembly is replaced, care must be used so these tiny leads don't get damaged. Breaking a wire would result in replacement of the entire drum assembly.

Video Playback Pre-amplifier

Because the playback signal coming off the video head is low voltage (less than one millivolt), amplification must be provided to increase the signal to a useable level for processing. This is the function of the pre-amplifier circuit. The pre-amplifier must provide gain while introducing as little noise as possible into the playback signal. Components are carefully selected to meet these gain low noise requirements.

The pre-amplifier is a tuned tank circuit. A tank circuit has frequency response characteristics that are determined by the values of its capacitor and inductor components. Transformers such as the rotary transformer and the windings and core material of the video head are inductors. Adding an adjustable capacitor to a transformer produces a tuned tank circuit. The tank circuit is tuned by adjusting the capacitor so the circuit resonates at a desired frequency. The pre-amplifier tank circuit is tuned to peak at 5 MHz so maximum output is recovered from the video heads during playback (especially at the upper end of the FM spectrum).

If the peaking circuit is adjusted with a capacitor and coil, the playback response characteristics of the tuned pre-amplifier circuit would be such that only those signals near the 5 MHz frequency would be correctly amplified.

A damping resistor is placed in the tuned circuit to dampen or reduce the amplitude of the 5 MHz peak and broaden the response characteristics of the tank circuit so all video frequencies are retrieved and amplified equally.

Adjustment of the pre-amplifier tuned circuit is made using a factory alignment tape that has an RF sweep waveform recorded on it. The RF sweep section of the alignment tape sweeps the frequency range of the entire video spectrum picked up by the video heads in 1/60 of a second. Fig. 6.17 shows the playback waveform pattern from a factory RF sweep alignment tape. Frequency markers are placed to identify important frequencies as the sweep range varies from zero to 5.1 MHz. Playback alignment of the tuned circuit is performed by adjusting the variable resistor control so it has no effect on the circuit. While watching the sweep waveform on the oscilloscope, the technician adjusts the variable capacitor so the peak of the tuned frequency is as outlined in the alignment instructions. This is normally between 4.5 and 5.1 MHz. The variable resistor is then adjusted to produce an RF sweep envelope with amplitude characteristics that are as equal as possible over all frequencies.

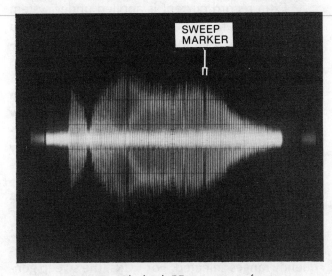

Fig. 6.17. Playback RF sweep waveform.

Proper video playback amplifier alignment is important to ensure that the playback characteristics of the video signal are equal for both video heads. Each video head has its own pre-amplifier and tuned circuit. If the pre-amplifier tuned circuit for one head is misadjusted, the FM portion of the signal which corresponds to peak white video in the deviation range would not be played back correctly. The result could be noise or streaks at peak white levels flickering at 30 Hz on the TV screen. This occurs when the video head pre-amplifier doesn't properly play back the FM frequency associated with peak white. Streaks or noise could flicker on the screen each time that particular video head attempted to playback peak white.

Many pre-amplifier circuits also permit an adjustment which balances the RF envelopes picked up and amplified by the pre-amplifier circuit providing equal playback RF signals from each pre-amplifier. Fig. 6.18 shows an unequal RF envelope. This VCR has no pre-amplifier balance adjustment. The imbalance is caused by a worn video head.

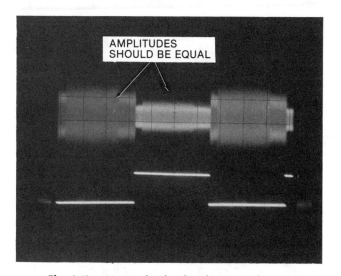

Fig. 6.18. Unequal video head output after pre-amplification.

The video head pre-amplifiers must be electronically turned on and off at appropriate times as shown in Fig. 6.19. When head A is playing back RF track A, the video head B pre-amplifier must be switched off. Video head B is not in contact with the tape most of the time that video head A is playing back a signal. If the pre-amplifier for video head B was on during head A playback, the B-head pre-amplifier would have no signal to amplify. Instead, the pre-amplifier would pick up external noise signals and send these to be demodulated along with the video head A FM signal, which would produce an extremely snowy picture.

Fig. 6.20 is a photograph of an oscilloscope presentation showing a pre-amp output and switching pulse.

A video head switching pulse is generated by the rotating video head drum. This signal is processed in the servo circuitry and will be described later. The important point to remember is that this head switching pulse is responsible for turning on the correct pre-amplifier at the appropriate time. Fig. 6.19 also shows the relationship between the RF signal output from head A, head B and the video head switching pulse.

Fig. 6.21 is a block diagram of Y playback processing. The pre-amplified FM signal is applied to a limiting circuit where the top and bottom portions of the FM playback signal are chopped off. Signal limiting eliminates noise and waveform distortion which can both occur at the positive and negative peaks of the FM signal. Any waveform distortion which occurs as a result of saturating the tape during record is also chopped off in the limiting process. Limiting provides the cleanest demodulated output possible. The limited signal is passed through a drop out compensator to reduce the unwanted visual effects that occur as a result of missing pieces of tape oxide.

DROP OUT COMPENSATION

A possibility exists that a few lines of video information could be missing during playback because of imperfections in the video tape. If the video tape is missing oxide particles from the tape being scratched or damaged, or the tape binder itself is missing or not strong enough to hold the oxide on the tape, a phenomenon known as *drop out* will occur. Drop out occurs when the playback RF signal is missing for short periods of time. Drop out is seen as tiny lines of snow which travel rapidly across the screen from left to right. The actual amount of drop out varies greatly, from less than a line to one or more lines. Generally, most video tapes have a limited number of drop outs and only one or two horizontal lines are affected.

The use of FM helps to reduce the sensitivity of the video playback processing circuits to variations in the playback FM signal. FM permits magnetic saturation of the video tape during recording. This allows the maximum amount of magnetic information to be retrieved from the tape in playback. Minor variations in playback amplitude levels do not affect the FM demodulator. This is be-

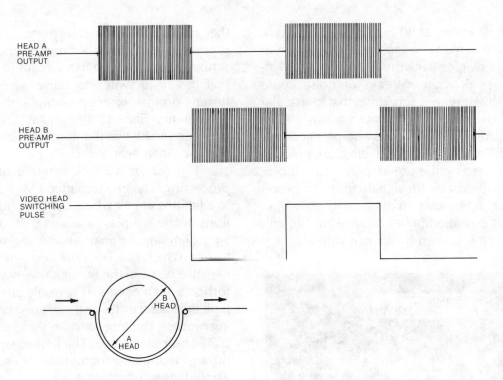

Fig. 6.19. The video head switching pulse turns on one pre-amp at a time.

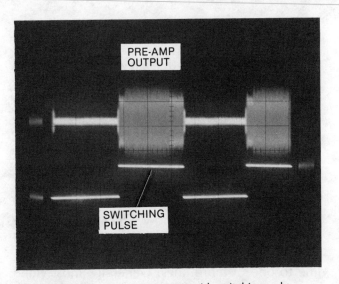

Fig. 6.20. Pre-amp output with switching pulse.

cause the peak-to-peak voltage of the FM signal is limited prior to demodulation. A limiting circuit is used to literally chop off the upper and lower portions of the playback FM signal. Minor variations in the level of the playback signal are literally ignored because the limiting circuit chops off and discards those portions of the signal anyway.

If a total drop out of the FM signal occurs, however, all video during the drop out is lost. A drop out compensator minimizes the visual effects of drop out on the TV screen. Drop out compensation (DOC) is a special circuit which identifies missing sections of RF information and inserts a good video signal from a previous line in its place.

Fig. 6.22 shows a block diagram of how a DOC operates. Playback RF from the limiting circuitry enters the DOC and is split in two directions. The first path for the RF information is through the main amplifier. For discussion purposes this will be considered to be a drop out-free RF envelope. After amplification, the signal is sent to the FM demodulator. This signal is also sent to a horizontal delay line which delays the signal by one horizontal line. Keep in mind that this good RF field contains 262-and-a-half horizontal lines. Each of these delayed horizontal lines is supplied to one of the inputs of the DOC switch waiting to be used. If it is not needed by the time the next line of information comes along it is replaced by the following good line of RF video information.

Now go back to the point where the limiter outputs the RF envelope. Imagine that the next vertical field of RF information comes along containing drop out. The peak detector circuit identi-

VCR Operating Theory   153

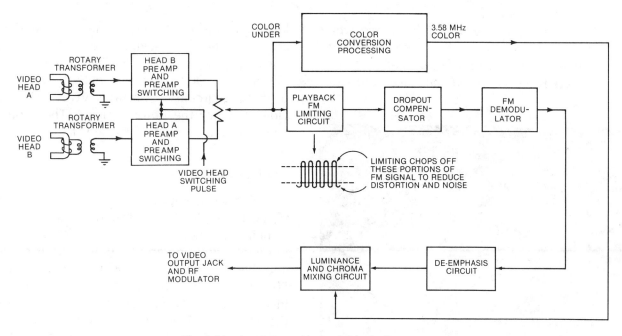

Fig. 6.21. Luminance playback block diagram.

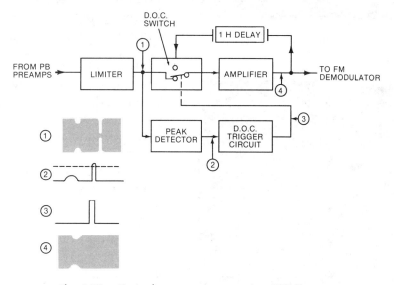

Fig. 6.22. Basic drop out compensation (DOC).

fies the exact point where RF energy is missing and sends out a pulse to the DOC trigger circuit. The DOC trigger circuit has a sensitivity level which has been adjusted in the alignment process to ignore minor fluctuations in RF energy levels and to send out a trigger pulse when RF information is missing. The DOC trigger pulse is output to the DOC switch while drop out is present. The DOC trigger switch toggles into the delayed RF position. The switch will remain in that position as long as the DOC trigger pulse is present. The DOC operates by injecting the previous good line of RF information and sending that on to the demodulator.

We see then the same horizontal line twice. Our eyes are not sensitive enough to observe this repeat action. Keep in mind that this single line of repeated video information is only one out of 525 lines. That's hardly enough for us to notice. If DOC were not to occur in the playback process, the result would be tiny visible lines of black and white snow rapidly moving across the screen which detract from the overall picture. Fig. 6.23 and Fig. 6.24 show video playback before and after drop out compensation.

Fig. 6.23. Video presentation before drop out compensation.

Fig. 6.24. Video presentation after drop out compensation.

It is interesting to note that drop out compensation works very well when there is only one line of horizontal information missing. Drop outs for longer than one horizontal line are also handled fairly well by the DOC. It is the last line of good horizontal video information which continues to be repeated over and over again, however. In this case, DOC forms a loop continually repeating the last good line of RF information over and over again. Practically speaking, drop out compensation is effective for only a few lines of missing video information. DOC occurs in only the luminance processing circuitry. Consumer VCRs do not provide DOC for the chroma signal. The drop out compensated, limited RF signal is fed into the FM demodulator circuitry.

FM Demodulation

The FM demodulator is based on the electronic principle demonstrated in Fig. 6.25. This principle recognizes that if a square wave signal of equal amplitude and time duration is rectified and filtered, the average DC level will be exactly 50 percent of the original square wave. If the positive going portion of that square wave maintains the same width or time duration and yet the pulses are placed closer together, the rectified and filtered DC voltage will increase because the negative going portion of the square wave is diminished.

Fig. 6.26 shows the basic operation for one FM demodulator. We should note here that this is only one example of how FM demodulation is accomplished. FM demodulation techniques vary not only between the different manufacturers but from one model to the next within the same manufacturer.

In our example in the figure, we see that limited FM enters a differentiator circuit that removes the DC component of the incoming square wave by converting the negative and positive going transitions of the square wave into sharp spikes. The positive-going spikes correspond to the positive-going portion of the incoming square wave; the negative-going spikes correspond to the negative-going portion of the original square wave signal. The frequency of the differentiated square wave is doubled in a doubling circuit. Diodes connected like those found in a common full-wave rectifier circuit can be used to perform this function. The doubling circuit inverts and adds the negative-going spikes to the positive-going spikes.

The positive-going spikes from the frequency doubler enter a monostable multivibrator (MMV) whose output is controlled by pulses from its incoming signal (in this case, the frequency doubler). The duration of the period of the MMV output

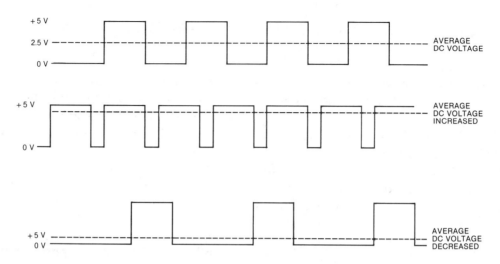

Fig. 6.25. Average DC levels vary with respect to the distance between square wave pulses.

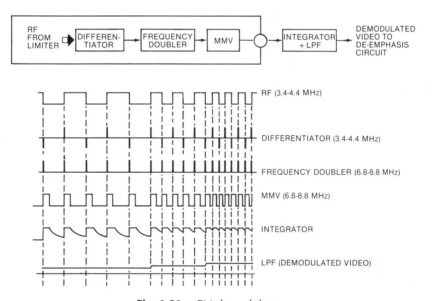

Fig. 6.26. FM demodulator.

pulse is controlled by external components such as resistors and capacitors. The positive-going portion of the MMV output is relatively short. As the incoming FM square wave bunches closer together representing high frequencies, the changes in the output voltage level of the MMV are also compressed. The output frequency has, in effect, been doubled. The output of the MMV is integrated and passed through a low pass filter to produce a DC voltage whose level directly corresponds to the frequency and period of the MMV.

VCR FM demodulators vary from one model to the next, however, they all share two main characteristics. First, they all double the frequency of the carrier in order to place the carrier ripple well above the video pass band frequency. Second, they act as pulse count or pulse density detectors. The higher the density of the FM signal, the higher the detected voltage output from the demodulator. Current FM demodulators are contained within ICs.

The demodulated Y signal now begins to closely resemble the original luminance portion of the recorded signal. Only two things need to be done to turn the signal back to its original form. During record pre-emphasis was applied to the luminance signal to preserve high frequency picture detail. This must be removed and the up-converted chroma must be added back into the luminance signal.

De-emphasis is performed on the video signal by first passing the demodulated signal through a low pass filter to eliminate any unwanted signal frequencies above the approximate 3 MHz range. De-emphasis is applied to the sync tip portion of the signal by clamping the negative-going excursions. De-emphasis of the video portion of the signal is accomplished by placing capacitors in the video circuit amplifiers within the de-emphasis block. The capacitors act to reduce the high frequency component of the signal slightly. The values of these capacitors are carefully selected to attenuate the unwanted pre-emphasized video spikes while maintaining proper high frequency video information. The de-emphasized video signal is mixed with the color up-converted chroma and applied to the video line output jacks and to the radio frequency (RF) modulator for channels 3 and 4.

E TO E

During the recording process, VCRs enable the video program being recorded to be displayed on the TV screen. This feature is known as *electronic-to-electronic* or E to E. E to E means that the incoming video signal is accepted by the machine and sent directly out to the video output jacks and RF modulator. This lets the operator simultaneously view and check the recorded signal. The E to E signal does not become an output after it has been completely processed in all machines. In consumer VCRs, E to E normally means that the signal coming into the VCR from the tuner or line input jacks is available at the RF modulator output or video and audio line output jacks after having gone through minimum signal processing. During the record mode, E to E is selected when the TV VCR button is placed in the VCR position. This causes the audio and video signals from the machine to be E to E signals.

SUPER BETA

Recently, Beta and VHS manufacturers discovered ways to improve the video picture reproduction by modifying and improving the video processing circuitry. The first manufacturer to introduce improved picture quality was Sony. They modified the FM circuitry to provide a way to play back a picture with 20 percent greater detail. They called this new procedure *SuperBeta*. SuperBeta circuitry shifts the FM deviation range from the original 3.5 to 4.8 MHz to a new 4.4 to 5.6 MHz deviation. 4.4 MHz corresponds to sync tip black while 5.6 MHz corresponds to 100 percent peak white. This increases the amount of room available for the lower sideband portion of the signal. Recall that much of the picture detail is contained within the FM sideband frequencies. With increased space now available for the lower FM sideband a greater portion of the lower sideband frequencies can be recorded onto the tape without interfering with the color-under signal. Beta manufacturers claim a 20 percent increase in picture detail over that obtained by the previous system. Fig. 6.27 shows the frequency allocations for the former Beta system and the current SuperBeta units with the SuperBeta switch on.

In some models, SuperBeta technology also provides a smaller video head core size which permits small guard bands between adjacent tracks in the Beta II mode. Because guard bands are now available in Beta II (previously they weren't), new de-emphasis circuitry is incorporated and improved noise cancellation and chroma crosstalk circuits are used. These new circuits provide picture detail and noise reduction never before achieved.

Beta manufacturers claim that the SuperBeta machines are basically compatible with the previous format. SuperBeta units have a switch which allows the SuperBeta machine to record and play back either standard or SuperBeta modes. Tests have shown that recording a picture in the SuperBeta mode and playing it back on an older conventional machine does play back relatively well. However, there can be some slight streaking in the high level luminance portion of the signal. This seems to be caused by overdriving the FM demodulator in the conventional Beta unit. This ef-

A. Former Beta system.

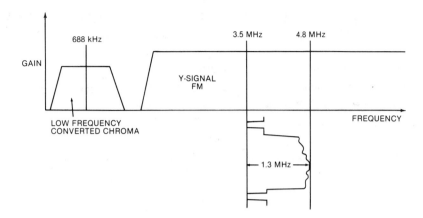

B. Current SuperBeta system with switch on.

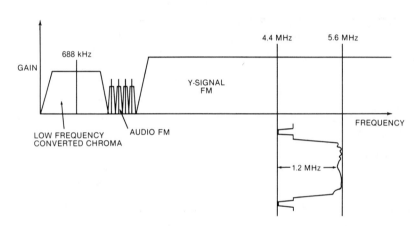

Fig. 6.27. A comparison of Beta and SuperBeta frequency allocations.

fect is minimal and the overall picture quality is very good. Excellent compatibility exists between conventional and SuperBeta units when the SuperBeta switch is off.

VHS HIGH QUALITY (HQ)

It didn't take long for the VHS manufacturers to also find a way to make slight picture quality improvements in their units. VHS manufacturers designate this improved picture quality as HQ, for *high quality*. VHS HQ does not modify the FM deviation frequency.

With VHS HQ, three areas of the video process were improved. First, the white clip level adjustment was modified to provide a 20 percent increase in the white clip level before the signal modulation. Second, luminance noise reduction circuitry was improved giving the appearance of crisper picture detail. Third, chroma noise reduction circuitry was improved.

Improvements in picture quality for VHS HQ are noticeable when comparing previous VHS units with those having the new HQ technology. Video resolution is not increased, but overall picture quality appears cleaner (due to noise reduction), and sharper (due to picture detail enhancement). HQ improvements are not as dramatic as those in the SuperBeta machines because SuperBeta improved both the video resolution and the noise reduction.

When dubbing or editing tapes on a VHS HQ or SuperBeta machine, follow carefully the manufacturer's suggestions concerning the placement of switches for noise reduction HQ and of the editing switches. Correct placement of the switches gives you the best picture quality of the dubbed tape. Noise reduction features provided in the HQ and SuperBeta units are especially designed for the best reproduction quality possible of playback recorded signal. These new noise reduction circuits change the characteristic of the frequency response during normal playback. The picture tube containing the TV screen responds best to video frequencies around the 1 MHz range. The signal output from some new VCR circuits emphasizes frequencies in the 1 MHz range as a way to improve the image on the screen.

However, this emphasis causes picture quality deterioration when a tape is being dubbed from one machine to another. The problem occurs because the FM modulator assumes the emphasized 1 MHz portion of the video signal is actual video information. The result is reduced quality of the edited tape. Some SuperBeta machines have an edit switch which configures the playback machine circuitry during the dubbing and gives the best quality picture. If you are using a VHS HQ unit, de-energize the noise reduction circuitry while you are editing. Otherwise the video quality of the edited tape will be very poor.

SERVO CONTROL

To properly play back video RF tracks, information must be recorded in precise locations on the tape and the VCR must be able to identify the location of this information. Therefore, RF tracks are identified to enable the machine to relate track A to video head A and track B to video head B. In addition, the speed and exact position of the video heads on the rotating video head drum must be carefully controlled. The video tape must be drawn through the machine at precise speed to ensure proper play back of sound and picture. Servo control circuitry maintains the correct head drum rotation and video tape speed.

The servo circuitry in a VCR has two main sections. Each of these sections has two subdivisions—the Video Head Drum Servo section and the Capstan Servo section. A subsection within each main section governs the coarse (free-run) speed of the capstan and drum. Another subsection, the Phase Control Unit, fine tunes the speed.

For example, the video head drum can rotate at a coarse speed of 1798 RPM, but the exact placement or position of the video head at any given instant in time must be accurately controlled. The free-run speed of the head drum and capstan is controlled by the Free Running Speed Section while exact placement of the video head is controlled by each Phase Control Unit.

EARLY BETA DRUM SERVO

Servo control of the phase and free run speed of the head drum and capstan is best understood by analyzing the operation of some of the early Beta VCR servos through the progression of VCR servo circuit technology to the complex servo circuits which exist today.

Early VCRs were driven by a hysteresis AC motor that rotated at a speed governed by the frequency of the AC power available at the electrical wall socket. Fig. 6.28 is a block diagram of a basic drum servo during record. Each of the two video heads record one vertical frame of picture information. Incoming vertical sync from the program being recorded is used to identify the vertical field.

During record, each video head must record its signal on the tape at a precise time. In addition, recording must be timed so that the vertical sync pulse is recorded about ten-and-a-half horizontal lines after the head comes in contact with the tape.

Vertical sync must be recorded at exactly the same position on the tape with each sweep of the head. To be sure this happens, the rotation and position of the head are closely controlled. During record, the video head drum must be speed-and-phase-locked to the incoming vertical sync signal.

The video head drum assemblies were belt driven by the hysteresis AC motor and connected to this motor by a series of belts and pulleys. The coarse speed of the video head drum assembly was, therefore, controlled directly by the speed of the AC motor. Pulley diameters were chosen so that the drum assembly rotated slightly faster than the AC motor.

The AC motor rotated at a precise 30 revolutions per second (1800 RPM). At this free-run speed, the video head passes over each track on the tape in slightly less than 1/60 of a second.

A color TV program is broadcast at a vertical field rate of slightly less than 1/60 of a second (1/59.94 sec). The drum servo uses the counter-electromotive force (EMF) of an electro-magnet applied to the shaft of the video drum assembly to

VCR Operating Theory   159

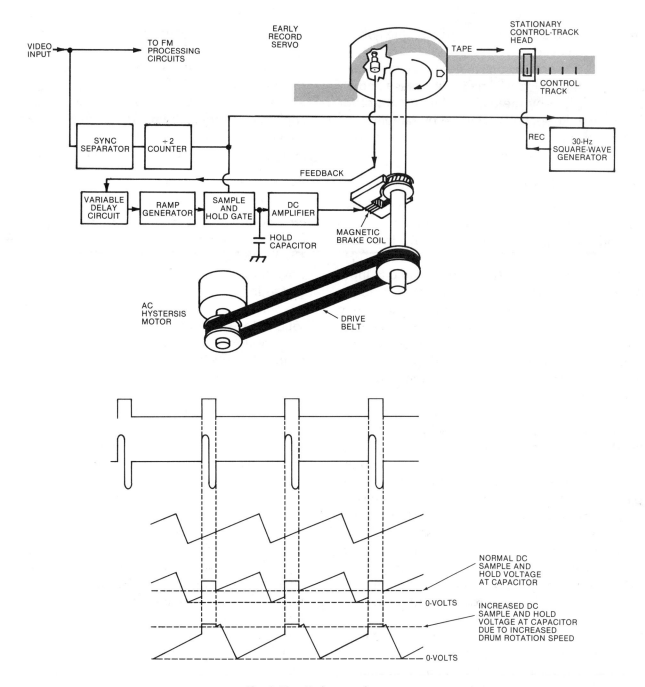

Fig. 6.28. Early record servo.

slow the rotation to a precise speed. The electro-magnet is energized by the servo circuit.

The electro-magnet counters the magnetic lines of force from a permanent magnet mounted on the rotating drum shaft. Opposing EMF slows the drum.

This drum servo circuitry first senses the position of the video head drum by monitoring the position of permanent magnets mounted on the rotating assembly. Next, the circuitry compares that position with vertical sync, then increases or decreases the amount of EMF to achieve the desired rotational velocity. If the head drum is rotating too fast (closer to its free-run speed), increased braking current passes through the magnetic braking coil and increases the applied EMF. If the video head drum is rotating too slow, less braking power is applied and the speed increases toward its free-run speed until the correct phase is achieved.

In this example, the free-run speed is controlled by the 60 Hz AC line input voltage and increased slightly by the pulleys. Phase control for fine tuning the speed is achieved by sampling the actual position of the video head drum and comparing it with incoming vertical sync.

Drum Record Phase

Proper record phase is achieved by separating the incoming vertical sync from the video signal and applying the sync signal to a divide-by-two circuit. The output of this circuit is a stable 30 Hz (29.97 Hz) square wave that is passed to one input of the sample-and-hold circuit. The 30 Hz square wave from the divide-by-two circuit is amplified and sent to the control track head where it is recorded on the extreme lower portion of the video tape. This recorded signal is called a *control track pulse* sequence and is used as a reference for playback servo operation.

On the video drum head assembly, two permanent magnets identify the exact position of the heads. As these magnets rotate on the head drum assembly, they pass a pick-up coil mounted on the stationary lower drum portion of the assembly. The pick-up coil senses the magnetic lines of force cutting across its windings and produces the series of pulses shown in Fig. 6.28. These pulses pass through an adjustable electronic delay circuit that compensates for small mechanical errors in the placement of magnets on the drum assembly. The magnets are aligned for proper phase timing during servo alignment.

The output of the variable delay circuit is fed to a ramp generator that converts the incoming delayed spikes from the stationary pick-up coil into a ramp DC voltage. The frequency of the DC ramp is controlled by the pulse stream from the stationary pick-up coil. As the rotation of the video head drum increases, the positive excursions of the ramp bunch together.

The sample-and-hold section compares the 30 Hz square wave from the divide-by-two circuit with the frequency of the ramp generator. The sample-and-hold circuit slices the ramp into 30 Hz sections, samples the voltage at the point where the square wave and ramp intersect, then passes the sampled voltage to a hold capacitor which charges up to the sampled voltage. The charge on the hold capacitor corresponds to the speed of the rotating drum assembly.

The sampled voltage on the hold capacitor is applied to a very high impedance (resistance) amplifier that increases the voltage and causes amplified current to flow through the break coil.

Early Beta Servo Error Correction

When the speed of the rotating head drum increases, the output frequency from the stationary pick-up coil and the ramp generator also increase. The incoming vertical sync, and the 30 Hz square wave remain constant during this period. The increase in the frequency of the ramp output from the generator causes the sample-and-hold 30 Hz square wave to intersect the ramp at a higher voltage position. This means that the DC ramp has changed position in comparison to the counted-down incoming vertical sync.

The hold capacitor charges to a higher voltage which, when amplified by the DC amplifier, causes increased current to flow through the brake coil. The larger current flow generates a stronger counter EMF which, when applied to the video head drum shaft, slows down the drum assembly. If the video head drum rotates too slowly, the DC ramp signals out of the ramp generator are farther apart. The 30 Hz square wave pulse samples the DC voltage on the ramp at a position corresponding to lower voltage which causes a lower voltage charge on the hold capacitor. The charge on the hold capacitor decreases current flow from the amplifier to the brake coil, less counter-EMF is applied to the rotating drum assembly, and the video head drum increases rotational velocity toward free-run speed until the correct phase is achieved.

Early Beta Record Capstan Servo

Early VCRs using drum servo control relied upon the speed of the AC motor and a series of belts and pulleys to control the rotating speed of the capstan shaft. The speed of the capstan shaft set the constant speed at which the video tape was pulled through the machine.

Video tape is placed between the capstan shaft and a rubber pinch roller. The pinch roller presses the video tape firmly against the capstan shaft causing the tape to move through the machine. Audio, reel-to-reel, and cassette tape recorders use a similar pinch roller and capstan technique to move the tape at precisely controlled speeds.

Early Beta Playback Servo Control

To properly play back recorded information, the video head must travel over the same area on the tape as it did during recording. This action is called *tracking*. Head A must travel over the head A RF track. If this doesn't occur, the video head will pickup noise from the video guard band or the adjacent RF track. With azimuth recording, head A doesn't recognize RF track B so the TV displays a screen full of snow or random video noise.

Several factors ensure that a video head travels over the correct RF track. First, the tape must be traveling at the same speed in playback as it did in record. Second, the video head drum assembly must rotate at the same speed during record and playback. Third, video head rotation must have the same phase relationship with vertical sync during record and playback.

Fig. 6.29 is a block diagram of the playback drum servo circuit. During playback, the speed of the video tape being drawn through the machine is controlled the same way it is during record. The AC motor, with speed determined by the frequency of the line voltage, controls the capstan through a series of pulleys. The free-run speed of the video head drum assembly is also controlled by the AC motor through belts and pulleys.

Vertical sync from the incoming video is no longer a valid reference for phase comparison in consumer machines, so it is disconnected from the drum servo assembly. Instead, the recorded control track pulses are used to compare the position of the video playback head with the vertical sync on the tape. The control track pulses are derived from the record signal vertical sync pulses and serve as the playback reference signal for the video head servo phase control.

Video Head Switching Pulse Formation

The current flow caused by the two stationary pick-up coil fields develops two voltage potentials on the inputs to a common flip-flop circuit. These two pulses trigger the flip-flop causing a 30 Hz square wave which is directly proportional to the position of the rotating video head drum assembly. This 30 Hz square wave becomes the video head switching pulses that turn on or off the appropriate video head pre-amplifier during playback. The 30 Hz head switching pulses are also used by the color processing circuitry during color-under phase modification for both VHS and Beta machines. The video head switching pulses are present during record and playback modes and are also used in the system control circuit.

The system control circuit monitors the video head switching pulse to be sure that the video head is rotating at the correct speed. If this circuit senses that the head is not rotating correctly, it will shut down the machine.

The video head drum position is sampled the same way in record and playback. The permanent magnets mounted on the rotating video head drum assembly pass across stationary pick-up coils generating head switching pulses and special pulses for the variable delay circuit and the ramp generator. The frequency of a DC voltage ramp generated by these pulses corresponds to that of the rotating video head drum assembly. This ramp signal is one input to the sample-and-hold circuit.

The second input to the sample-and-hold circuit comes from the pulses picked up by the control track head. The stationary control track head is mounted in the longitudinal audio head stack assembly. The pulses picked off the tape by the control head are no longer square waves but are pulses with pointed positive- and negative-going spikes. These spikes represent the positive- and negative-going portions of the recorded square wave signal. Recall from Chapter 5 that video tape can't be magnetized unless the voltage across the recording head coils is changing. The only time that the voltage is changing in a square wave is at the leading or trailing edge of that square pulse. This explains why the original square wave signal appears so distorted when played back.

The playback control track signal is amplified and applied to the sample-and-hold circuitry. This signal is used to sample the DC voltage on the point where the positive-going spike intersects the DC ramp as shown in Fig. 6.30. The negative-going portion of the spike is not used. The sample of this DC ramp voltage is applied to the DC amplifier and

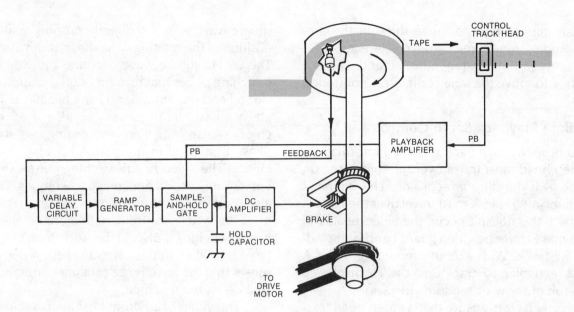

Fig. 6.29. Early playback servo.

from there to the brake. Phase control occurs in a similar manner to that of the record servo circuit except that the control track pulses instead of the record vertical sync are used as the 30 Hz reference.

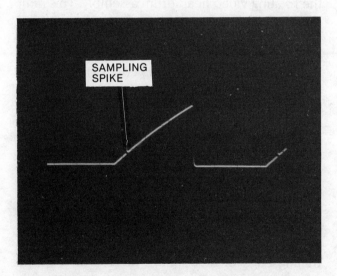

Fig. 6-30. The DC ramp is sampled where the spike intersects.

NEW SERVOS FOR MULTIPLE-SPEED MACHINES

The servo control operations covered so far worked very well for the first Sony Beta Max units because these machines recorded and played back in a single speed. The introduction of two-speed machines created a need for capstan servos which could handle multiple tape speeds. The single speed system previously described was modified by Sony to produce Beta I and Beta II machines. The AC motor is still used to control the rotating video head drum assembly, and the head drum servo circuitry is nearly identical to that described. However, a separate capstan motor was required to permit proper tape speed control. In the meantime, some VHS manufacturers introduced VCRs with direct drive drum motors rather than the belt-driven motors common to Beta machines. Many early VHS units had two-speed record and playback capability. With the introduction of multiple speed machines and direct drive motors, servo processing circuitry changed, although the principles remained the same. Both drum and capstan servos still are speed-and-phase controlled. The incoming vertical sync acts as a reference for phase control during record, and the control track pulses act as a reference for during playback. The position of the video head drum must be continuously monitored for phase comparison during record and playback.

Fig. 6.31 is a speed and phase servo processing block diagram for a direct drive head drum motor.

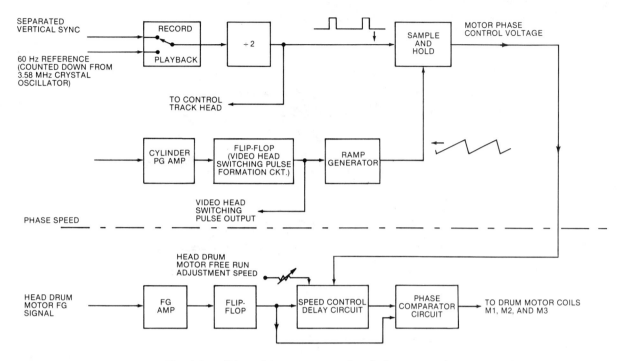

Fig. 6.31. Direct drive motor speed and phase control.

Direct Drive Motors

The direct drive video head drum motor is a DC motor whose speed and video head phase position are controlled by the incoming voltage. Since control is so important, many VCRs have a DC brushless motor. The brushless motor is ideal because of its durability and ability to be accurately controlled.

The video head drum motor consists of three main coils with three poles for each coil as shown in Fig. 6.32. The coils are mounted on the stationary portion of the video head drum assembly. The motor flywheel is attached to the rotating shaft of the motor. Three permanently magnetized individual ring magnets are mounted on the flywheel. Each of the three ring magnets performs individual functions. The two top magnets provide feedback to the servo circuitry for speed and phase control. The inside magnet is closest to the stationary coils and is part of the motor.

The inside ring magnet consists of twelve individual poles alternating between north and south as shown in Fig. 6.33. This inner ring magnet is positioned close to the stationary main coils in the motor. The main coils are turned on and off individually by switching transistors inside the drum motor integrated circuit (IC). As these main coils are energized, magnetic lines of force cause the magnetic flywheel mounted on the video head drum shaft to rotate. The three poles on each coil are mounted 120 degrees apart so that, by energizing the three poles of each coil in the correct sequence, each set of poles will take its turn in forcing the magnetic flywheel to rotate. As the motor rotates, each of the three main coils are energized 120 degrees after the previous main coil.

Direct Drive Motor Speed and Phase Control

Switching the main drive coil drive transistors is performed inside the drum motor IC. In order for the transistors to turn off each of the respective main coils at the correct time, the position of the video head drum flywheel must be identified. This is a function of the two upper permanent magnetic rings on the rotating flywheel.

The magnetic poles are mounted on the upper portion of the flywheel in such a way that they identify the exact position of the flywheel. Fig. 6.33 shows the magnetic polarity of these magnetic rings. As the flywheel rotates it passes across two Hall-effect IC's (called *Hall ICs*) which are per-

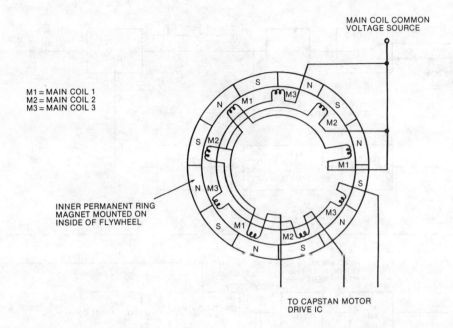

Fig. 6.32. Direct drive drum motor has three main coils with three poles each.

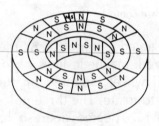

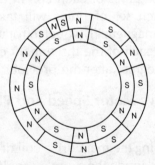

Fig. 6.33. Head drum flywheel has three ring magnets.

manently mounted on the stationary part of the lower drum assembly.

A Hall IC is an integrated circuit that is sensitive to the lines of force from a magnetic field. It acts like a switch and changes state each time the field of a magnetic pole is reversed. Fig. 6.34 shows the nine poles and a Hall IC after the flywheel has been removed.

Two Hall ICs are mounted on the video head

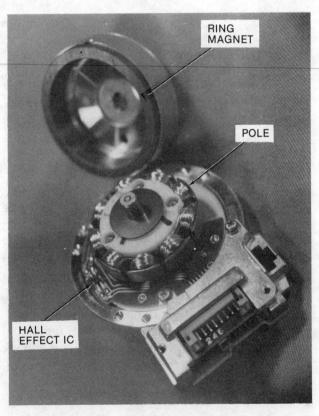

Fig. 6.34. A dismantled head drum motor.

drum motor. One of these ICs detects the magnetic information from the outside upper ring on the flywheel. The second IC detects the magnetic information on the inner ring of the upper portion of the flywheel. The two Hall ICs produce pulses which correspond to the exact position of the rotating head drum motor at any given moment. This information is returned to the drum motor drive IC where it is processed and used for switching the main coil drive transistors. Fig. 6.35 shows the output of the inside and outside ring Hall ICs as well as the derivation of the electronic switching transistor pulses.

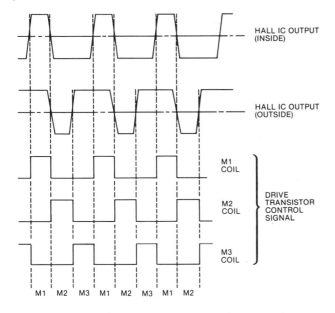

Fig. 6.35. Hall effect output signals.

You may recall that the outside upper ring shown in Fig. 6.33 has two magnetic poles that are much narrower than the other poles on the magnet. These two poles identify the position of the video head on the rotating drum assembly. This information is used to generate video head switching pulses.

Information from the Hall ICs on the video head drum motor is processed inside the IC as shown in Fig. 6.36. This signal not only switches the motor main coil drive transistors, it generates *PG pulses* (pulse generator) and *FG* (frequency generator) *pulses*. PG pulses which are like the pulses which come off of the stationary pick-up coil in Fig. 6.28 identify the position of the video head drum. FG pulses, created by the upper inside and outside flywheel permanent magnets, represent the speed at which the motor is rotating. FG and PG information exits the Drum Motor Drive IC

and are processed by another IC to produce the video head switching pulses.

While the motor is rotating at 1800 RPM, the FG pulses occur at the rate of 180 Hz. The relationship between the two Hall IC outputs, the 30 Hz PG signal, the 180 Hz FG signal, and the video head switching pulse is shown in Fig. 6.37.

FG and PG signals are used by the video head drum servo to control both phase and speed of the motor.

Servo Circuits for Direct Drive Motors

Fig. 6.36 shows that one of the inputs to the Video Head Drum Motor Drive IC is a torque control voltage. This voltage comes from the video head drum motor servo circuit and controls both the speed and phase of the motor. Digital processing techniques derive the servo control voltages for the drum and capstan drive ICs.

To understand this technique, it is helpful to consider the servo technology used in earlier direct drive servo controlled motors. Fig. 6.31 (shown earlier), a block diagram of an early drum servo, is divided into two basic sections. The lower portion describes the free-run speed circuitry, and the top part describes the phase control circuitry.

The free-run speed of the video head drum motor is controlled by the FG signal coming from the flywheel on the video head drum motor. Since the incoming FG signal is too weak to be processed, it's amplified before it goes to a flip-flop. The flip-flop converts incoming FG to a 50/50 duty cycle pulse (50 percent positive, 50 percent negative square wave) which is fed into one part of a phase comparator and into a speed control delay circuit. A change in speed is detected by the phase comparator as it compares the 50/50 duty cycle from the flip-flop with the pulses coming from the speed control delay circuit. A change in speed represents a change in the phase of the two pulses. An output control voltage from the phase comparator to the drum motor coils causes an appropriate speed adjustment. The free-run speed section of the servo is capable of adjusting free-run speed errors. Fine adjustment is controlled by the phase section of the servo. The servo speed controller keeps the video head rotating at approximately 1800 RPM. Speed control in the drum servo is

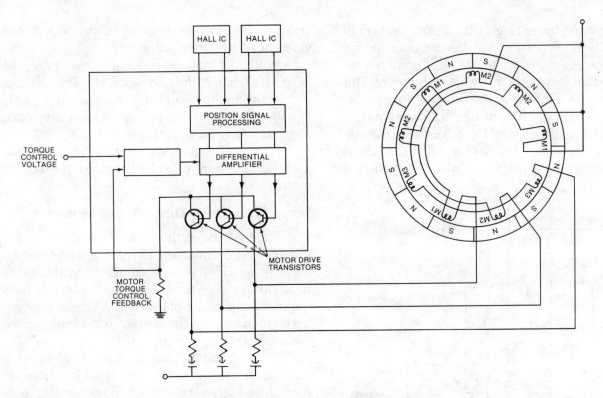

Fig. 6.36. Drum motor and drive electronics.

achieved in the same manner for both record and playback modes.

Drum motor phase control for both record and playback is similar to that shown in Fig. 6.28 and Fig. 6.29. During record, the video head drum position is identified using PG pulses. The position of vertical sync and the incoming record video signal is identified by separating out vertical sync, dividing vertical sync by two (now 30 Hz) and comparing the two. Separated vertical sync is converted into a square wave and fed to a sample-and-hold circuit. Cylinder PG is transformed into a DC ramp signal that is also fed into the sample-and-hold circuit. A DC voltage representing the exact place where the 30 Hz square wave intersects the DC ramp is held by a small hold capacitor, amplified, and sent out to mix with the free-run speed control circuitry.

A slight phase error in the rotating drum speed creates PG pulses which bunch if the speed increases or which separate if the speed decreases. The end result is that the DC ramps occur closer together or farther apart. In the sample-and-hold circuit, the positive-going 30 Hz pulse is placed at a position higher or lower on the ramp corresponding to an increase or decrease in the detected phase error. An increase in frequency corresponds to an increase in the voltage output from the sample-and-hold circuit. After this error voltage has been mixed with the free-run speed drive voltage, it is applied to the video head drum motor to correct both coarse and fine motor speed errors.

In record, the 30 Hz square wave derived from separated vertical sync is also applied to the control track head and recorded on the tape just as it was in early Beta units.

SEPARATE HEAD DRUM AND CAPSTAN SERVO RESPONSIBILITIES DEFINED

During record, the video head drum servo ensures that each video head arrives over the track at the right time. If each head contacts the tape at exactly 10 horizontal lines prior to vertical sync as referenced by the incoming video record signal, the head drum servo is performing correctly.

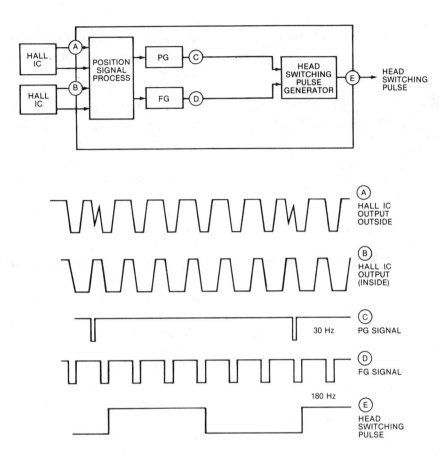

Fig. 6.37. Head switching pulse formation.

Capstan Servo

The capstan servo keeps the video tape traveling through the machine at a steady speed. The servo for a DC controlled capstan motor uses capstan generated FG pulses and video head drum generated PG pulses to correct for speed and phase errors. The capstan servo circuit in multiple speed units must be designed to handle different record and playback speeds. PG pulses arriving from the video head drum motor occur at a constant frequency, while the capstan FG pulses vary in frequency as different record and playback speeds are selected.

The capstan servo circuitry performs three major functions. First, it ensures that the video tape is drawn through the machine at a constant speed. Second, it makes sure that the recorded RF tracks on the tape arrive at the video head drum at the correct time. The third function of the capstan servo circuit involves the record and playback speeds of the video tape itself.

During record, the capstan servo pulls the tape through the machine at the speed selected by the operator. In playback the capstan servo circuitry monitors the speed of the recorded tape and automatically changes the video tape speed for proper playback. If, during the recording mode the video tape speed select button is changed, the playback capstan servo circuitry monitors the new speed and adjusts the tape speed for proper playback.

The capstan motor has a permanent magnet attached to the flywheel that produces an FG signal. Two direct drive capstan motors and one belt drive capstan motor are shown in Fig. 6.38.

Fig. 6.39 is a block diagram of an early VCR capstan servo. A close relationship between the capstan servo and the video head drum servo must be maintained during the record and playback process. To enable this relationship, video head drum PG pulses are used to form a DC ramp in the phase comparison circuitry for both the drum and

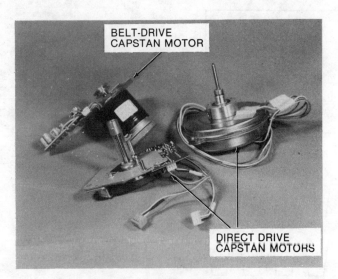

Fig. 6.38. Capstan drum motors.

capstan. If the video head drum motor slows down slightly, the capstan servo can make minor corrections and slow down tape movement.

Fig. 6.39 shows incoming FG pulses going through a divider circuit that produces a 240 Hz signal. (FG frequencies vary according to manufacturing designs. This frequency is used for discussion only). The divider circuit is key to the record and playback capstan speed selection circuitry. It divides by one, two, or three as directed by the incoming divider control circuit (to be outlined shortly).

The 240 Hz signal is sent to the capstan speed control delay circuit which functions in the same manner as a similar circuit for the drum motor. The FG signal is first converted by a flip-flop to form a 50/50 duty cycle square wave. The output of the flip-flop is fed into the speed control delay device and also routed to the input of a phase comparator. Speed corrections are made by comparing the delayed 50/50 duty cycle with the phase of the non-delayed signal. Speed errors are detected by the phase comparator and a capstan control DC voltage is applied to the capstan motor.

The 240 Hz counted-down FG signal from the FG divider circuit is also applied to a divide-by-eight circuit to form a 30 Hz output signal for the pulse shaping circuit. The 30 Hz signal is shaped for processing and is applied to one input of the sample-and-hold circuit. PG pulses from the video head drum motor are applied to the same ramp generator circuitry used for the video head drum phase control. The output ramp signal is also ap-

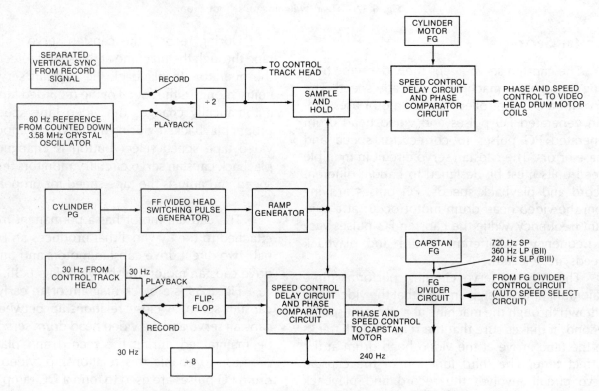

Fig. 6.39. Capstan and drum servo circuitry work closely together.

plied to the capstan sample-and-hold circuit. The DC voltage corresponding to the point where the 30 Hz signal intersects the DC ramp is held by a hold capacitor and applied to the speed control delay circuit for capstan phase correction.

When the capstan rotation speed changes, phase and speed control are identical with the exception of the operation of the divider circuit at the capstan FG input point. When the capstan is rotating at its fastest speed (SP, Beta I), the capstan FG frequency is 720 Hz. LP or Beta II speeds generate 360 Hz FG pulses. SLP or Beta III speeds generate 240 Hz FG pulses. Except for this difference of speed, the record capstan servo function is the same.

In record, the divide-by circuitry is controlled by an incoming DC signal that is selected on the front panel of the VCR. The record speed determines the DC voltages at the input control points of the FG divide circuit. In playback, the divider is controlled by the auto speed selection circuitry. The FG divide circuit operates the same in playback as it does in record except the DC control voltages controlling the divider come from the auto speed selector.

Capstan Playback Servo

The capstan servo control works the same in playback as it does in record with two exceptions. First, the FG divider circuit is no longer controlled by the record speed selection switch on the front of the VCR but is now controlled automatically by an auto speed selection circuit. The second difference is the source of the 30 Hz square wave. In record, the source of this signal came from FG pulses which had been divided down. In playback the source of the 30 Hz signal is from the control track head.

Control Track

The early, one-speed Beta units used the 30 Hz control track signal in playback for phase correction of the video head drum assembly. (See Fig. 6.29.) This method was used because the capstan speed was directly controlled by the rotation of the AC motor. The AC motor speed is directly proportional to the frequency of the AC line voltage.

VCRs now use the capstan controlled pulse to phase lock the capstan motor rather than the video head drum. The 30 Hz control track signal from the tape identifies the exact position of the vertical sync pulse as it is recorded. An earlier section described how the capstan servo made sure that each video head RF track reaches the video head drum at the exact time that the corresponding head begins to sweep across the tape in playback. The head drum and capstan servos must work closely to accomplish this task. The responsibility of the head drum servo is to make sure that the video head contacts the tape at precisely 10 horizontal lines prior to vertical sync. The capstan servo must make sure that the video RF track is there at the same time that the head arrives.

In playback, the head drum servo stabilizes the rotating speed and phase of the video head drum while the capstan servo advances or retards the speed of the tape slightly to match the appropriate RF track with the correct head. If the capstan servo loses phase lock, bands of snow such as that shown in Fig. 6.40 drift through the picture. This display looks initially like a tracking problem, but with tracking errors, the snow band does not drift.

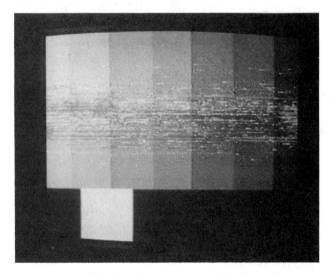

Fig. 6.40. Drifting band of snow caused by loss of capstan servo phase lock.

A capstan servo problem can be confirmed by monitoring the output RF waveform from the video head pre-amplifier and switching pulse. A display like that in Fig. 6.41 definitely confirms that the capstan servo requires alignment or repair.

Control track pulses identify the location of the RF tracks on the tape. This is why the capstan

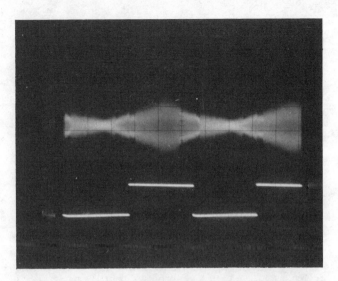

Fig. 6.41. Display confirming a capstan servo problem.

servo must use the control track as a playback reference. The control track pulse from the tape is shaped to perform the sample-and-hold operation. PG pulses from the video head drum are converted into a DC ramp in the ramp generator.

This is the same ramp generator used for the video head drum phase circuit. The ramp pulse is applied to the sample-and-hold circuit where it is compared with the control track square wave pulse stream. The voltage output of the sample-and-hold circuit corresponds to the DC voltage where the 30 Hz square wave and ramp signal intersect. The sample-and-hold voltage acts like a phase control that is added to the free-run speed control delay circuitry for fine adjustment of the capstan speed.

Tracking Control

Tracking control, on the outside of the machine, applies a control voltage to the speed control delay circuit and fine tunes the capstan servo phase control during playback. The tracking control has a center detent position to physically indicate the setting for normal operation. This adjustment enables fine tuning of the capstan servo control electronics to correct for tracking differences between different machines. Ideally, all machines will play back correctly with the tracking control in the center detent position. If for some reason yours doesn't, the tracking adjustment will help correct this error.

The tracking control advances or retards the phase of the capstan servo by altering the relationship of the FG and PG pulses. The tracking control is turned clockwise or counter-clockwise from the center detent position to slightly advance or retard the phase relationship of these two pulses and to obtain the best playback picture. Advancing the phase of the capstan servo with the tracking control causes the tape to kick forward a little so the RF tracks on the tape are positioned properly for detection by the video head. Retarding the phase of the capstan servo with the tracking control causes the video tape to slow down briefly and allows the video heads to catch up with the RF tracks on the tape. The tracking control, then, adjusts the phase relationship between the rotating video head drum assembly and the RF information recorded on the tape by altering the phase of the capstan motor.

Auto Speed Selection

In playback, automatic speed selection is performed using the capstan servo and the electronics outlined in Fig. 6.42. The FG divider in the capstan circuit acts on the incoming FG signal to produce a 240 Hz FG signal at its output regardless of capstan speed. To accomplish this task, the FG divider must know how to divide the incoming FG signal. This information comes from the auto speed select circuitry that monitors the incoming control track pulses by measuring the distance between each of the pulses. Two voltages from the auto speed select circuit directly correspond to the speed of the tape.

Assume that, during record the first five minutes of tape are recorded in LP speed. While still recording the TV program, the record speed selection switch is accidentally changed to the SP position. The tape is recorded at SP speed until suddenly you realize that the program you are recording is longer than you anticipated so you quickly turn the record speed selection knob to the SLP position.

After completing the recording session the tape is rewound and played back. The part of the program recorded in LP is played back first and control track pulses occur at exactly 30 Hz. The VCR playback continues at this speed until the point is reached where the recorded speed changes to SP. This causes the SP control track pulses to occur farther apart since the machine is

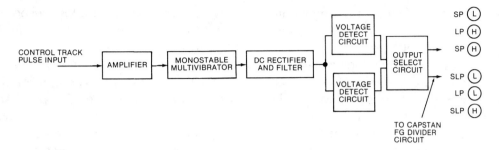

Fig. 6.42. Auto speed select circuit.

still playing back in LP. Actually, the speed is doubled, so the distance between the pulses is twice what it should be (15 Hz).

The auto speed select circuit forms square wave pulses using a *monostable multivibrator* (MMV). The MMV generates a short duration square wave every time it receives an incoming control track pulse. The farther apart the incoming control track pulses, the greater period of time that passes before the MMV generates another positive-going pulse. These pulses are filtered and converted into a DC voltage.

Because these pulses are farther apart, the average DC filter voltage is lower. This voltage level triggers two voltage detection circuits designed to change output voltage in proportion to the input voltages they sense. The output voltages toggle between a high logic state of about five volts and a low logic state of zero volts (logic high or logic low). The respective output pulses are passed to the capstan FG divider circuit which is preprogrammed to divide incoming FG signals by a predetermined amount as selected by the logic high and logic low levels from the auto speed select circuit.

Next, in this playback scenario, the point is reached where the recorded speed changes from SP to SLP. The control track pulses were recorded in SLP but are being played back in SP so now they occur at a 90 Hz rate rather than the normal 30 Hz. The MMV generates a 90 Hz signal which is filtered and converted into a higher than normal DC level. The auto speed select voltage detectors modify their outputs to a logic state which allows the FG divider circuit to function properly.

While these speed changes are taking place, the capstan free-run circuitry must make dramatic voltage changes to adjust the capstan motor to the right speed. This is done using the free-run speed control portion of the capstan circuitry. The speed control delay circuit senses the error in the incoming 240 Hz divided-down FG signal and makes the necessary corrections to adjust the capstan motor to the correct course speed.

Head Drum Playback Servo

In playback, the video head drum free-run speed and phase control operates essentially the same as it did during record with one major difference. During record, phase control used incoming vertical sync as a reference. During playback (in consumer machines), the separated vertical sync signal is not used. Instead, an internal 60 Hz reference is selected. This 60 Hz signal comes from the 3.58 MHz oscillator in the chroma circuitry. It is counted down to provide the 60 Hz reference for playback. An accurate 3.58 MHz oscillator is important for proper chroma record and playback, and it plays an essential part in playback drum servo phase control. The incoming 60 Hz reference was shown in the upper left hand corner of Fig. 6.31. This signal is divided by two and processed the same as the incoming vertical sync was during record.

Digital Servo Processing

Servo circuits in current VCRs use digital technology for phase and speed control. Digital processing in modern VCR servo sections is contained almost entirely within ICs. ICs are not serviceable parts. They must be replaced when they become defective. Digital servo processing still works much the same as the analog processing in older machines. The same reference PG and FG signals are used and the PG, FG, 60 Hz reference, vertical sync and CTL pulses are still required. The difference lies in how these signals are used within the servo ICs to generate the drum and capstan speed and phase control voltages.

Fig. 6.43 and Fig. 6.44 are block diagrams describing the digital drum servo and capstan servo circuitry. Within the figures are two blocks labelled *Pulse Width Modulator* (PWM). This is a square wave oscillator whose period varies as controlled by an input correction signal. Fig. 6.45 shows the output of the PWM during correct and incorrect rotation speeds of the motor being controlled. When the motor is rotating at the correct speed, the PWM output is a 50/50 duty cycle. If the motor is rotating faster than normal, the PWM signal outputs a square wave whose positive-going period has a longer duration. If the motor is rotating slower than it should, the positive period of the square wave has a shorter duration. Fig. 6.46 is a PWM waveform as viewed on an oscilloscope. These square wave PWM signals are filtered and converted into a DC control voltage. The amount of DC voltage produced by the filtered PWM signal corresponds to the length of time the positive portion of the signal is present. The faster the motor rotates, the higher the DC voltage produced. The opposite is true if the motor is rotating too slow.

The use of PWM in producing phase and speed comparisons may appear similar to the methods used in the speed control delay circuit and phase comparator as shown in Fig. 6.31. This is partially true, but digital technology is more complex and, yet, much more accurate than analog. Earlier servo technology (like that previously described) required a lot of alignment to maintain proper performance. Digitally processed servo sections have only one or two adjustments available; in some cases, no adjustments are provided.

However, this new technology responds rapidly to phase and speed errors and provides highly reliable performance.

A good example of the improvements associated with digital servo processing is apparent when you compare the automatic speed selection circuitry of current and early model VCRs. During record, if the speed select switch on an early model VCR is changed, and then the tape is played back, the changes in speed are quite visible on the TV screen. The video appears to be moving at the wrong speed and the audio track sounds misaligned with the talking figures. This lasts for a second or two until the machine stabilizes at the new speed.

On a VCR built using current digital processing technology for the drum and phase servo control, the capstan and head drum servos operate so precisely that no disturbance is noticeable on the screen or in the audio. The capstan motor and video head drum compensate so rapidly that only close scrutiny will detect the speed changes.

Special Effects

VCRs featuring special playback effects, such as still frame, pause, and slow motion playback, place special demands on the servo circuitry. The capstan servo, in response to commands from the system control circuitry, performs most of these special effects.

When playback pause is engaged, the capstan motor must be stopped at a position which provides the best playback signal from the recorded tape. Some machines stop the tape with the capstan motor and allow the system control circuits to sample the RF envelope coming from the video heads. If the system control circuitry de-

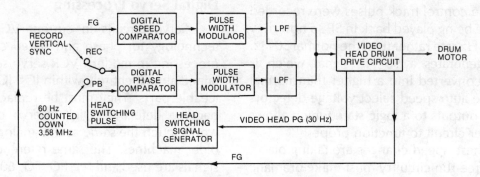

Fig. 6.43. Digital drum servo block diagram.

VCR Operating Theory   173

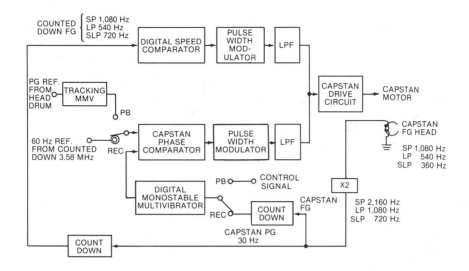

Fig. 6.44. Digital capstan servo block diagram.

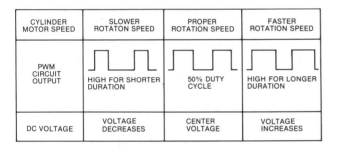

Fig. 6.45. Drum speed control output.

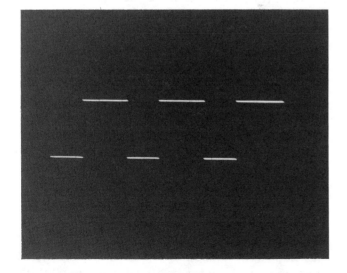

Fig. 6.46. Oscilloscope presentation of PWM.

termines that the RF envelope is insufficient to produce a good still frame picture, it commands the capstan motor to pulse forward until a good RF envelope is detected. The system control circuitry provides most of the commands for the full electromechanical operation of the VCR. Functions and features of the system control will be described shortly.

Slow motion forward and reverse speeds are also controlled by the system control unit. System Control directs the capstan servo to jog the tape in the forward or reverse direction to provide good slow motion video playback. In many machines slow motion playback causes rapid starting and stopping of the capstan motor. The tape is stopped for an instant, a video frame is generated on the screen, and then the tape is jogged forward or in reverse until the next frame can be displayed.

System Control

The System Control circuitry is responsible for coordinating all of the electronic and mechanical functions of the unit. Performing many of the functions of a simple computer, this circuitry monitors the status of the machine, acknowledges and responds to commands from the operator, and initiates instructions for correct operation.

System Control turns on and monitors the operation of the power supply. In most VCRs, when the power button is pressed, power is not actually supplied until System Control has checked the status and condition of the electronic and mechanical sections. Once satisfied that all is well, System Control generates command signals to the power supply and causes the complete unit to be energized.

When a VCR is turned off, System Control is still active. Power off does not eliminate all voltages within the machine. Instead, the VCR shifts to a power-down stand-by mode waiting for the next power-on instruction. Power consumption is extremely low when the VCR power switch is in the OFF position, but some voltages are still present inside.

System Control is comprised of several ICs that work together to monitor and operate the VCR. The heart of System Control is a microprocessor integrated circuit. The microprocessor constantly monitors the electronic and mechanical operation of the machine, receiving status signals and generating commands that are sent throughout the machine. As the brain of the VCR, the System Control microprocessor (there may be several) makes decisions and coordinates the operation of all parts of the unit. Acting much like a busy taxi cab radio dispatcher, it receives incoming requests from the operating control buttons and switches on the outside of the VCR chassis, checks the current status and operating mode, makes appropriate decisions, and sends command instructions to the VCR electronic and mechanical subsystems. If it senses an incorrect or improper condition, the microprocessor generates a stop command telling the machine to cease operating or to shut down until the situation can be resolved.

The microprocessor controls the power on and off sequence. When you depress the power button, you do not have direct control over the on and off functions. Instead, your action instructs the microprocessor to begin the power up (or power down) sequence.

The microprocessor accepts operator inputs from the front panel such as Play, Rewind, Stop, Record, etc., and issues commands to the electronic and mechanical sections of the unit. When record is selected, the microprocessor checks the status of the record safety tab switch located within the VCR. If the record tab has been removed from the video cassette, the record safety tab switch will activate causing the microprocessor to prevent recording on the tape.

The microprocessor also changes operating states. If the VCR is placed in the rewind mode and the play button is pressed during tape rewinding, typical microprocessors are programmed to know that a stop command must be issued before initiating the play command. The microprocessor responds as if you had first depressed the Stop button and then depressed Play.

Many VCRs offer Cue and Review (search forward/search reverse) features that are activated by depressing the Fast Forward or Rewind buttons in the play mode. The microprocessor can be programmed to recognize the difference between Cue and Review and the actual rewind and fast forward modes.

It also controls many special effects functions. The servo section described how the microprocessor monitors the RF envelope coming from the video heads and issues appropriate commands to the capstan circuitry to provide proper still frame pictures when Play Pause is selected.

The System Control microprocessor also determines the direction of the reel table motor rotation. The two reel tables insert slightly into the bottom of a video cassette where the cassette tape hubs are located. The reel tables are driven by pulleys called *tires* via a reel table motor. In the case of direct drive reel tables, each reel table has its own individual motor as shown in Fig. 6.47. There are no belts or pulleys to drive these reel tables. System control determines the direction and speed of the reel table motors. In special effects playback, a command for slow motion in the forward or reverse direction changes the motion of the reel table.

In units with direct drive reel table motors, System Control monitors the tape tension caused by the reel table motors as the tape is drawn around the head drum assembly. The video tape must move around the head under constant tension. VCRs without direct drive reel table motors use other means to control tape tension.

The microprocessor also controls tape load and unload functions. It prevents the video tape cassette tray from ejecting the tape until the tape has been fully unloaded and retracted back into the cassette. A motor and gear arrangement are used, as shown in Fig. 6.48, in front load VCRs to lower the cassette onto the VCR reel tables shown in Fig. 6.49. System Control senses when a video tape cassette has been placed into the cassette bas-

Fig. 6.47. Direct drive reel motor with one hub removed.

Fig. 6.49. The cassette is lowered onto reel tables.

ket. It causes the tape cassette to be drawn into the machine and lowers the cassette into the correct operating position. The microprocessor in System Control prevents other mechanical operations until the cassette has been correctly positioned.

Fig. 6.48. Front load mechanism.

The microprocessor functions closely with the tuner and clock timer circuitry. The automatic timer IC operates an internal clock that notifies the microprocessor when a preprogrammed time has arrived so auto recording can begin. It notifies the microprocessor that it should begin recording and also selects the preprogrammed channel.

Separate microprocessors can be used to control the timer and the tuner. The timer and tuner microprocessors work together to produce lighted displays on the front panel of the VCR. These digital displays can include the time of day, a tape counter, and the mode of operation currently selected. On VCRs having a fully electronic tuner (no mechanical moving parts) a microprocessor controls channel changing. The operator cannot change channels on an electronic tuner while recording. System Control acts to prevent changing a channel until the unit is placed in Record/Pause or Stop. This control was programmed into the microprocessor to avoid accidental channel changes if you unknowingly brush against the channel change button.

The System Control circuitry also accepts commands from a remote control transmitter. Circuitry within System Control receives and amplifies incoming commands from a hand-held remote control. This information enters the microprocessor where it is translated into appropriate internal commands for the VCR circuitry.

Whether the selected mode of operation enters System Control from a remote control transmitter or from the front panel function buttons, the microprocessor accepts the commands and takes appropriate action.

System Control is responsible for engaging the pinch roller against the capstan after the tape has been threaded in the machine. It controls the engaging and disengaging of the brakes which lock the reel tables in place in the Stop mode. Many gears, pulleys and levers must be moved into correct position to perform tape thread, unthread, normal play, fast forward, rewind and the other functions. Through a series of switches and optical sensing devices, the microprocessor monitors the mechanical state of each major portion of the VCR. These switches and sensing devices send DC voltage-level changes, or (in some cases) digital pulses to the microprocessor as state identifiers.

System Control regulates automatic features that are often taken for granted during machine use. When a video tape is fast forwarded or rewound, the microprocessor senses the end of the tape and immediately stops tape movement. The video tape has a special leader on it for identifying the start or end of the tape. In a VHS machine, these leaders are clear plastic. In Beta tapes, leaders are made of metal foil.

End of tape sensing in VHS units is performed using a small incandescent light or an infrared light emitting diode (LED). This light sticks up into a recess near the tape door of a cassette that has been inserted into a VCR. To the right and left of the cassette tape are two light sensitive photo detectors. When either end of the tape is reached, the light inserted inside the cassette housing shines through the clear leader onto one of the photo detect devices. The photo detector receiving the light generates a pulse to the microprocessor signifying that the end of the tape has been reached.

In Play, Record or Fast Forward, the tape travels from the left hand spool to the right hand spool as viewed from the front of the VCR. When the end of the tape has been reached, the clear leader is on the left side of the cassette tape. Light shining from the LED is detected by the photo sensor on this side of the chassis. Some VCRs perform an automatic rewind function upon sensing the end of tape.

During Rewind, whether initiated automatically or manually, the tape travels from the right hand spool to the left hand spool. When the end of the tape is reached, light shines through the clear leader at the beginning of the tape on the left spool and is detected by the right hand photo sensor. The photo detector output is interpreted by the microprocessor and a stop command is generated causing immediate cessation of tape movement. This happens so rapidly that the tape stops within two inches after an end-of-tape signal is produced.

Beta units also have end-of-tape detection at both ends of the tape although the leaders are constructed using metal foil. In this case, the end of the tape is recognized by small metal detectors on the left and right hand side of the chassis cassette housing positioned very close to the actual tape.

Memory stop is another feature controlled by the microprocessor in most machines. The operator can identify a particular place on the video tape by zeroing the tape counter, and depressing the memory button. During rewind, the microprocessor initiates a stop command when all zeroes are detected on the tape counter. Some VCRs have sophisticated tape counters that not only display a number representing tape usage, but that also display the amount of tape remaining. This is achieved by detecting the speed of the two rotating reel tables and comparing their different rotation rates. FG signals from each of the reel tables are sent to the microprocessor where they are compared. A short microprocessor program calculates the amount of tape used. Tape remaining indicators have demonstrated accuracy to within a few minutes.

The major function of System Control is to monitor the condition and operation of the machine. If abnormal operation is detected, an auto stop or emergency stop command is generated causing the VCR to enter the stop mode and in some cases power down. In this role, System Control monitors such things as reel table position, capstan and drum rotation, moisture condensation (moisture will cause the tape to stick to the drum), the amount of time required to thread and unthread the tape, and, in VHS machines, the end-of-tape sensor lamp. If any of these functions are sensed to be incorrect, an immediate stop is initiated.

The microprocessor also controls the Pause limit function. If the machine is left in Pause for an extended period of time, a possibility exists that the video tape may be damaged because video tape motion is held, and the tape is pulled against the rotating video head drum. The high speed rotating heads create friction causing thermal stress in the tape. This heat may cause the oxide binder to break down allowing the oxide coating to come off severely damaging the tape. Some tapes have been found with abrasion diagonals so severe that no oxide remained leaving only clear backing. The microprocessor monitors the amount of time that the machine is left in Pause during Record and Playback. If a specified time is exceeded (usually around five minutes) the microprocessor places the machine in Stop, or it releases Pause and the VCR resumes the mode it was in when Pause was initiated.

System Control design and application methods vary between manufacturers. Each attempts to be unique in its application of the microprocessor. Custom microprocessor ICs are common and their operation often varies substantially between models from the same manufacturer.

UNDERSTANDING MICROPROCESSORS

Knowledge of logic and digital processing is helpful in understanding the operation of System Control. System Control methods vary between VCRs, and diagnosing the information communicated within the system control circuitry requires sophisticated truth tables, so this discussion will be limited to a general overview using block diagram explanations and general principles.

Digital systems are inherently more reliable than mechanical systems with moving parts. System Control failure generally occurs in mechanical switches, but secondary sources for problems include electronic sensing devices, poor solder joints, and plug contacts.

Fig. 6.50 is an overall block diagram of the System Control circuitry. All mechanical and electronic functions are monitored and controlled by this circuit. As shown, sometimes several microprocessors are used to interpret data and generate commands.

Fig. 6.51 and Fig. 6.52 describe multiple microprocessor configurations. Fig. 6.51 shows the communication paths for several microprocessors working in conjunction. The control communication between the system blocks is called *handshaking*, an appropriate word to describe the close relationship and mutual communication which occurs in multiple microprocessor designs. Data lines

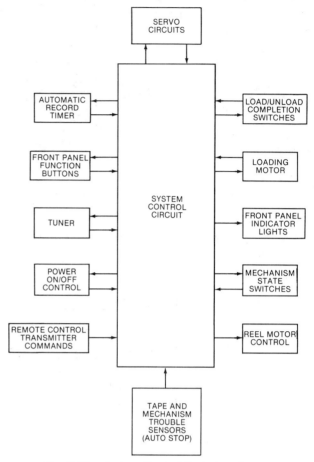

Fig. 6.50. System Control basic functions.

are used to pass primary information between the microprocessors under control of Ready and Acknowledge handshaking signals between the microprocessor that is ready to receive data and the microprocessor transmitting data. The Ready/Acknowledge line is also used to inform the transmitting microprocessor that the information has been received. This acknowledge signal is a shift in the output voltage level.

Some microprocessors use an input-output expander (I/O expander) to enhance their capabilities. With the introduction of many additional VCR features, System Control responsibilities have increased. To handle the added functions, it was necessary to either increase the number of input and output pins on the main microprocessor chip or to develop other means for receiving instructions and issuing commands such as interleaved input/output where data is passed in serial streams (multiplexed) or is exchanged on bidirectional pins.

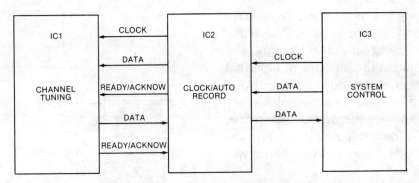

Fig. 6.51. System control handshaking.

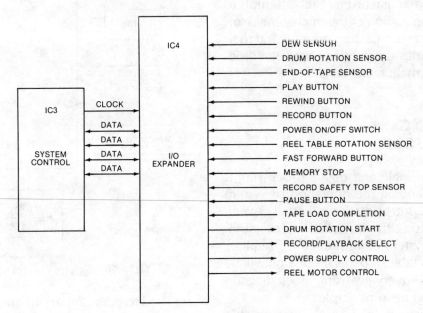

Fig. 6.52. Input/output expander.

The I/O expander is designed to accept additional information from many sources through pins on the IC and then to convert this information into data that can be transmitted to the main microprocessor on a few lines or circuit traces. The microprocessor can also send instructions to the I/O expander on a few lines while the commands are received and retransmitted out on many pins. Bidirectional data is passed to and from the main system control microprocessor via four data lines as shown in Fig. 6.52.

A pair of microprocessors requires several things in common before they can communicate back and forth. They must be able to receive and transmit data which they both can understand. They must be connected by some common means, and they must be initialized or reset to a ready state when power is first applied so they can receive and transmit their first instructions. Reset is performed each time power is applied.

Digital information is transferred over data lines as binary waveforms or digital pulse trains. Microprocessors use a common high-frequency oscillation called a *clock* to move the data on the data lines and to open or close logic gates which interpret and process the data. The frequency of the serial clock is determined by the designer and the type of microprocessor used. Typically these frequencies are in the one to two MHz range. Fig. 6.53 shows the relationship between the serial clock and the serial data pulses. The serial clock pulses are used to translate the serial data pulse trains.

Fig. 6.54 is a block diagram of a typical micro-

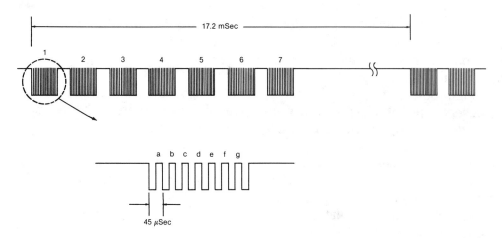

Fig. 6.53. Serial clock data pulse train.

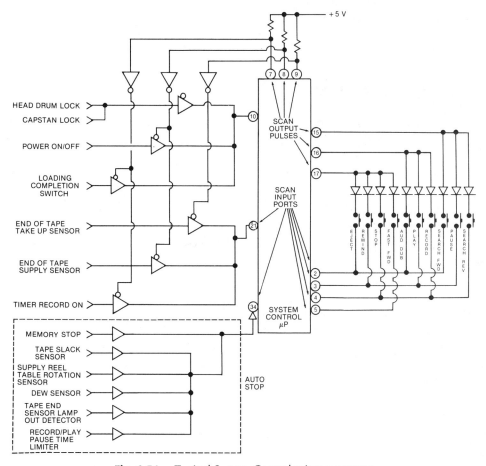

Fig. 6.54. Typical System Control microprocessor.

processor. As shown, the microprocessor is a single IC with many input and output pins. The number of pins can vary, but usually VCR microprocessors have between 40 and 64 pins. The microprocessor produces output pulses called *scan pulses* that are transmitted out through pins or ports to various switches and sensing circuits. When a switch closes, or makes contact, the scan pulses are returned back to the microprocessor through an input port. The microprocessor is wired in the circuit so it knows which function is being requested or which sensor switch is activated. Using a code scheme, the microprocessor can recognize the output scan pulse that is being received by an input port. In this way, the microprocessor can accept different scan pulses in a single input port, and let four inputs and four outputs provide multiple control functions.

System control seems complicated at first because it has a lot of activities to control. But, the process is simple if you remember that system control must receive input (status) information that it can produce outputs that control the individual functions in the unit. Put another way, system control is a circuit design using ICs that must be prodded, pushed, and clocked into and through each task being performed.

AUDIO RECORDING TECHNIQUES

Until recently, VCRs used only one method for recording audio onto the tape. This method was identical to that used in conventional audio recording techniques. The incoming signal was amplified and equalized to overcome the undesirable frequency characteristics of the audio head during playback and recorded on the tape using a stationary audio head. Recently Sony introduced an audio recording method using the rotating video heads mounted on the head drum. The conventional audio record method is a standard feature on both Beta and VHS machines, while the more recent method is an option. VHS and Beta manufacturers use different methods and have different names for the optional sound reproductions. Beta manufacturers call their method *Beta Hi-Fi*. VHS manufacturers call theirs *HD* for *High Definition* audio. For compatibility with non-hi-fi, non-HD units, all VCRs today still provide conventional stationary audio recording and playback, even if they include the new audio feature.

Manufacturers call the two audio record and playback methods *linear audio* and *audio FM* (AFM). Linear audio is stored on a tape using a stationary audio head which records a linear track on a section of the tape reserved for the audio head. AFM is recorded on a tape using audio heads on a rotating video head drum. These audio heads record the signal on the tape using designated FM frequencies.

Linear Audio Recording

Monaural audio was the only option available until stereo audio was developed. Monaural has also been called *mono* or *single channel* audio. Stereo audio uses the same recording techniques as monaural recording and playback but has two separate channels for left and right (Channel 1, Channel 2) audio. The video tape space designated for the audio information was divided in half enabling two audio tracks to occupy what had previously been the space of one track (Fig. 5.40 and 5.41).

The following description applies to monaural linear audio systems. VCRs with stereo linear audio are identical in principle to the monaural units except they have two identical audio record and playback circuits. In linear audio machines, the stationary audio head has two sections in the same space as the entire monaural audio head for recording the left and right channels.

Fig. 6.55 is a block diagram of linear audio recording. Incoming record audio enters the audio processing circuitry through a common line. Audio from the tuner, an external microphone, or the line level input jack is applied to a common connector at the input to the audio amplifying circuitry. Typical audio levels from the tuner section reach the audio circuit at standard line level voltages of around 0.7 V peak-to-peak (p-p), as does audio from an external source entering through the Audio In or Line In jack in the VCR. VCRs can also have an external microphone input. Typical microphone (mic) levels range from 1.0 to 10 mV p-p. Tuner audio and line input audio levels are attenuated (reduced) to mic level by resistor networks prior to the first audio amplifier. This normalizes all inputs to the first amplification stage to mic level regardless of the source.

Audio into the audio processsing circuitry is selected by a series of switches. The switch in the microphone jack gives priority to the mic input. When a microphone is plugged into the jack, line and tuner audio sources are disconnected. The line and tuner switch shown in the figure routes the audio signals to the processing circuitry from the selected source. Many VCRs don't have an ex-

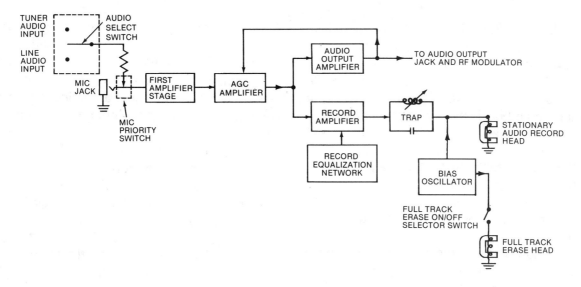

Fig. 6.55. Conventional audio record block diagram.

ternal switch for selecting line or tuner inputs. The switching automatically occurs inside the Line Input jack when a plug is inserted.

Incoming mic level audio to the first audio amplifier is amplified and sent in three directions. The first direction is into the audio AGC (automatic gain control) which samples the level of the amplified audio and automatically adjusts the volume level for optimum recording. Audio AGC acts like an automatic volume control to reduce the amplification if levels are too high and to increase amplification if levels are too low. Typical audio AGC is not as sophisticated as AGC and ACC in the chroma and video processing circuitry. If incoming audio has long periods of very quiet sound, AGC will attempt to increase those levels. Programs containing very loud audio will have the signal peaks limited by AGC.

The second direction that amplified audio takes is to the audio output circuitry. This line level audio out is passed to the Audio Output jack and the Channel 3 and 4 RF Modulator for E-to-E audio monitoring.

The third audio signal direction is into the record amplifier. The record amplifier increases the audio level and passes the signals to the audio record playback head. The record amplifier has equalization amplified to correct for the nonlinear playback response of the audio head as described in Chapter 5. Equalization is performed using discrete resistors, capacitors, and inductors in the amplification process causing different amplifier response to various frequencies. In this case, low frequencies are amplified more than high frequencies.

The audio from the record amplifier passes through a bias trap that is placed in the circuit to stop high frequency audio bias from being fed back into the record amplifier. The audio bias oscillator operates at a frequency of about 60 kHz. Enough of the bias oscillator is mixed with the audio during record to cause operation in the linear region of the hysteresis curve as explained in Chapter 5.

Bias is also applied simultaneously to the full track erase head and to the audio erase head during normal audio recording. The full erase track erase head eliminates all audio, video, and control track signals from the tape during record using a high frequency bias oscillation signal. Bias fed to the audio erase head eliminates previously recorded audio from the tape.

The switch for the full track erase head eliminates head bias during insert audio and insert video recording modes because it's not good to turn on the full track erase head during these functions.

Linear Playback Audio

In linear playback, audio information is read off a tape using the stationary audio head. The low (about one millivolt p-p) audio signal is amplified and equalized as shown in Fig. 6.56. The equalization amplifier boosts the playback signal producing an opposite effect to the incoming frequency record equalization amplifier. The output from the equalization amplifier is passed to an audio output amplifier which boosts the audio signal to a level suitable for external use via the Audio Line Output jack and the Channel 3, 4 RF Modulator.

VHS VCRs with the stereo linear audio feature may also have the capability to rerecord Channel 1 audio (right channel) after the original audio and video program has been stored on the tape. This form of audio insert editing is provided to give the operator two separate audio channels for special applications. For example, in videotaping a special family event, the original audio program could be recorded on Channel 2 while background music could be recorded on Channel 1 without interfering with the Channel 2 audio or the recorded video. During playback, the two channels of audio could be mixed together producing the original audio with dubbed-in background music. In these type VCRs, Channel 2 is used to record the main or primary information because of its placement at the very upper edge of the tape. Because the Channel 1 audio track is placed close to the outer edge of the tape, it is subject to minor variations in playback level. Such variations are caused by minor tape edge damage from improper tape handling by the VCR tape guides and mechanism. Therefore, the Channel 2 main audio is more protected than the Channel 1 dubbed-in audio. Professional editors use this same channel allocation method when dubbing-in audio to be mixed with the main part of a program.

Linear Audio Specifications

Linear audio record and playback performs well for most applications. Specifications are published by VCR manufacturers that describe the audio, record, and playback limitations of each machine. These specifications reflect three major areas of audio reproduction concern: (1) harmonic distortion, (2) frequency response, and (3) wow and flutter. Harmonic distortion is a measurement of signal deformity associated with normal recording, playback, and signal amplification. In distortion, the original signal is not perfectly reproduced during playback. For example, if a perfectly shaped sine wave is recorded on a tape, that same sine wave may appear normal during playback, but when compared on an oscilloscope with the original signal some minor variations may be visible. This is called *harmonic distortion*. It occurs because frequencies generated in record and playback produce small amplitude harmonics that combine to distort the original signal.

The measurement of *frequency response* determines the upper and lower frequency range for the machine. Writing speed limits the audio frequency response. Linear track audio frequency response typically ranges between 50 Hz and 10 kHz. Slow speed audio recording reduces the upper frequency limit. Beta 3 and SLP frequencies rarely extend above 6 kHz. Lower frequency response specifications are not affected by recording speed.

Audio *wow and flutter* occur when the video tape is not pulled through the machine at a steady speed. This distortion affects the original recording signal and is noticeable as a change in the pitch of constant tones. If a single note is recorded on a tape and sustained for several seconds, wow and flutter are heard during playback as a slight change in pitch. *Wow*, a measurement of the rate of change of pitch, received its name from the sound produced by the slow changing of the pitch of a tone. *Flutter* measures the fast rate of change in pitch of a constant tone. The rapidly changing audio pitch produces a fluttering sound. Both wow and flutter are results of video tape not being drawn through a machine at steady speed. Both wow and flutter can be present during playback.

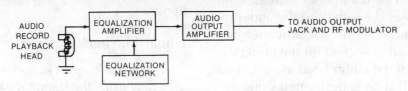

Fig. 6.56. Linear playback audio.

Wow and flutter are not changes in volume They are changes in pitch or frequency. Imagine a 1 kHz tone recorded on a tape without wow or flutter and then played back on a VCR with a mechanical misalignment causing slight variations in playback tape speed. The ear would hear the 1 kHz change in frequency. This change is directly related to the variation in playback speed. As the tape is drawn faster through the machine, the pitch frequency shifts upward. As the tape movement slows down, the playback pitch decreases. The rate at which this occurs differentiates between wow and flutter. Flutter occurs many, many times a second. Wow is a much slower shift in frequency.

Wow and flutter prevention is less effective as the tape speed is reduced during Record because it's increasingly difficult for the mechanical parts of the VCR to steadily pull the tape through the machine as tape speed is reduced. Unless specifically stated otherwise, published specifications for VCR wow and flutter are defined for the SP or Beta II modes only. Most machines have wow and flutter that can be detected at the slowest recording speeds. The best audio recording occurs during faster recording speeds. Frequency response and wow and flutter are optimized when the audio is recorded in standard play or Beta II speed.

Audio Frequency Modulation (AFM)

Sony engineers discovered that it was possible to frequency modulate a recorded audio signal using frequencies between the color-under signal value and the lower luminance sideband cutoff. Fig. 6.57 shows the typical bandpass frequency. The space between the color-under signal with its sidebands and the lower Y FM sideband cutoff represents available frequency space where an AFM signal could be injected. By applying the left and right incoming audio signals to two separate modulators oscillating in this available frequency slot, an audio FM signal could be recorded and played back. Sony found it was possible to use the rotating video heads to record this AFM signal. The AFM information is added to the color-under chroma and the Y FM signal, and then recorded on the tape.

Extra room is required between the Y FM lower sideband cutoff and the 688 kHz color-under signal with its associated upper sidebands because the video heads cannot discriminate be-

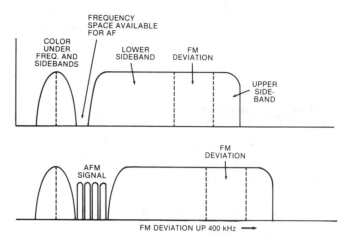

Fig. 6.57. Beta Hi-Fi shifts the FM deviation frequency up 400 kHz to allow extra room for the AFM signal to be recorded between the lower sideband cutoff and the color-under frequency.

tween the extreme lower sideband video information and the AFM signal whose frequencies overlap slightly. Elimination of the overlap is necessary to prevent video interference. The AFM and the Y FM lower sidebands are separated by shifting the Y FM frequency up 400 kHz. This provided the frequency space for AFM signal injection and complete signal interchangeability between all Beta units. If the frequency of the Y FM were not increased, the lower sideband would require chopping at its upper end. The result would be a loss of some high frequency detail. In reality, the color-under frequency upper sideband cannot be reduced because this would produce more loss in color detail. Moving the Y FM up 400 kHz was the practical solution to eliminate interference.

In playback, the video heads pick up adjacent track audio signals. The AFM signal is modulated at approximately 1.5 MHz, and the plus and minus six degree video head azimuth angle difference isn't sufficient to reduce adjacent RF track crosstalk at these low frequencies. Adjacent channel crosstalk during playback is eliminated in the Beta units by highly selective band pass filters and the use of four separate record modulation frequencies.

Chroma crosstalk cancellation techniques can't be used for audio signals because audio information doesn't contain the same phase relationships. Chroma crosstalk cancellation is based on the fact that each adjacent line maintains a specific phase relationship with the previous line. This is not true for audio, because audio signals maintain no phase relationships with each other.

AFM RECORDING

Fig. 6.58 shows the modulation and recording of AFM signals. Incoming audio for both the left and right channels enters from the left side of the figure.

Left channel audio enters into the frequency modulator where it's converted to a 1.53 MHz FM signal that is passed through to the coils of video head B as left channel audio. While head B is recording, both left and right channel audio are present so the right channel must also be recorded by that head. The source for the head B right channel information is from incoming right channel audio that enters the 1.68 MHz frequency modulator where it is split and applied in two directions. One direction is through the head A coil which isn't in contact with the tape at this time. The other signal path leads into one input of a balanced mixer. Another input to the balanced mixer is a 150 kHz sine wave generated by a crystal oscillator. The two frequencies are added in the balanced mixer producing sum and difference frequencies. The 1.83 MHz sum of the frequencies is allowed to pass through the band pass filter and on to head B as right channel audio.

Why Four Frequencies?

Left channel information is applied directly to the B head from the main modulator at 1.53 MHz. Both left and right channel audio information must be recorded on the tape at the time each head is in contact with the tape. If the AFM frequency for the left channel were recorded by both the A and B heads without frequency alteration, adjacent channel crosstalk would not be eliminated, because both RF tracks contain some of the same frequencies. To prevent this situation, the 1.68 MHz right channel AFM signal is converted by a mixer circuit to a higher frequency before being applied to the B head. This heterodyne frequency converter mixer is called a *balanced modulator*.

AFM recording through the A head begins when right channel audio information enters the main frequency modulator and is applied to the A head at 1.68 MHz. The left channel audio signal is also recorded on the tape through the A head. This left channel AFM has been frequency modulated to 1.53 MHz in the main modulator and applied to a balanced mixer. Like the mixer for head B, this mixer also receives a 150 kHz sine wave signal from an internal oscillator. The balanced mixer output is the sum and difference of 1.53 MHz and 150 kHz. These signals are applied to a band pass filter which allows only the 1.38 MHz difference frequency to pass. The 1.38 MHz AFM

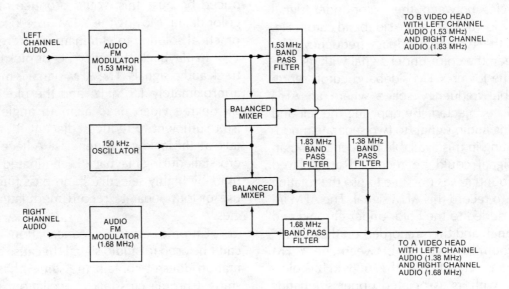

Fig. 6.58. Beta AFM record.

signal is mixed with the 1.68 MHz carrier signal and applied to the A head.

Playback

Unlike the B head frequencies, the A head AFM frequencies are different for the left and right channels because different band pass filters are used so only select frequencies can pass. Unwanted frequencies, such as those from adjacent channel crosstalk are rejected by the band pass filters, and only the desired frequencies are passed.

Fig. 6.59 is a block diagram of Beta AFM playback processing. The band pass filters separate left and right channel audio information for both the A and B heads.

In playback, AFM frequencies from the A video head are applied to balanced mixers (modulators) that are heterodyned frequency converters and shift the frequency up by 150 kHz. These frequencies then equal those coming from the B head. The 1.53 MHz information from the A and B heads represents the left channel audio information. The 1.83 MHz information from the A and B heads represents right channel information.

The 1.83 MHz and 1.53 MHz AFM signals from both heads are applied to a video head switching circuit. The electronic switches are energized by the video head switching pulse and select the correct input from a video head over which the tape is passing. The signals from the head not in contact with the tape are prevented from reaching the FM modulator so noise won't be present in the audio signal.

Head Switching

The AFM signal is applied to head switching circuitry that is more complex than that for the video head. Video head switching is performed at a point on the TV screen six-and-a-half lines before vertical sync. This is not seen by the observer be-

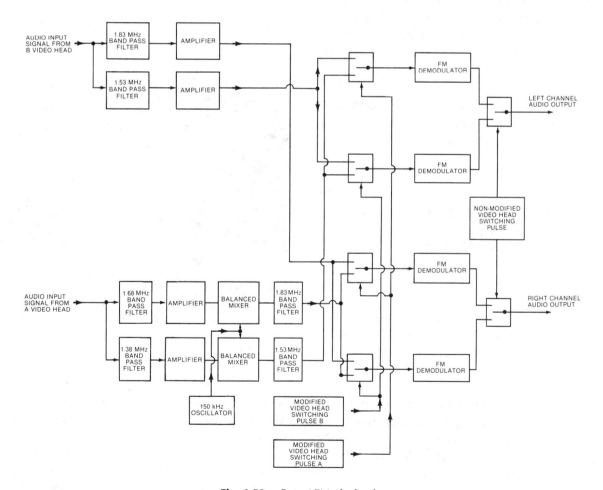

Fig. 6.59. Beta AFM playback.

cause switching occurs below the normal viewing portion of the TV image. Video head switching causes interference in the TV picture but it occurs in an unseen portion of the screen.

With AFM head switching, however, there is no convenient time to switch and keep the scan point below the viewing level of the TV observer. Audio is continually available at the speakers so head switching for AFM must be handled carefully to prevent audible popping noises. Audio interference is aural not visual. For this reason, AFM head switching is performed once before and once after the AFM demodulators in Beta systems.

As shown in the block diagram, head switching is performed prior to demodulation by a series of four switches. The head switching pulses which toggle the four AFM head switches derive from the video head switching input signal. The AFM signal is applied to four separate FM demodulators. Two FM demodulators are used for the 1.83 MHz signal, and two are used for the 1.53 MHz AFM signal. Each output from the FM demodulators is applied to a second head switching circuit. A fine head switching pulse is then applied to switch the demodulated AFM. Careful timing of the AFM head switches, both before and after AFM demodulators, produces a continuous stereo audio signal from the rotating video heads without any detectable popping sounds.

The audio output signal from the AFM demodulators and final switching circuitry is applied to a sample-and-hold circuit that eliminates audio dropout resulting from missing RF information on the video tape. AFM audio dropout occurs simultaneously with video dropout. The sample-and-hold circuit detects and holds an average audio signal level. The DC level represents the amplitude of the audio signal at any given period of time. If dropout occurs, this representative DC voltage is injected into the audio signal and fills the audio voids created by RF dropout with the average DC level.

Beta units can shift the Y FM carrier up by 400 kHz because of their writing speed. VHS machines use a slower writing speed which limits their ability to shift the Y FM signal. So VHS designers developed another technique to record and play back AFM signals.

VHS High Definition (HD)

VHS HD audio requires a separate pair of heads specifically designed for AFM recording. These heads are mounted 180 degrees apart and 120 degrees ahead of the adjacent video head as shown in Fig. 6.60. Fig. 6.61 is a VHS video head drum with four video heads and two AFM heads. The head gaps on these audio heads are placed at an angle of plus or minus 30 degrees. This large offset angle provides adequate crosstalk isolation from adjacent RF tracks during playback.

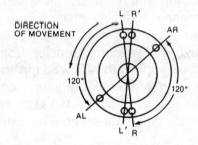

Fig. 6.60. The audio head precedes the video head by 120 degrees.

Fig. 6.61. Six head VHS head drum assembly.

VHS AFM frequencies are 1.3 MHz and 1.7 MHz for the left and right channels respectively. Mounting the AFM head gaps 60 degrees relative to each other provides adequate audio RF track crosstalk isolation for low AFM frequencies. This also provides suitable isolation between the Y FM and AFM signals which occupy the same RF track.

The difference in azimuth between the A video head and the A audio head provides good isolation between the Y FM and AFM signals. This isolation allows the Y FM lower sidebands and the AFM signals to occupy the same space in the frequency spectrum without causing picture interference.

Fig. 6.62 shows the relationship of the video recording frequency spectrum and the position of the AFM signal.

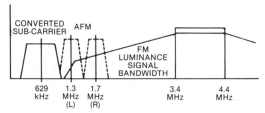

Fig. 6.62. AFM position within the audio-video frequency spectrum.

The AFM signal occurs in the lower sideband portion of the Y FM frequency range. In Beta systems, AFM and Y FM are isolated and do not occur in the same general frequency band.

Depth Multiplexing

One VHS recording technique which differs from that for Beta AFM places the audio signal in the video FM lower sideband without causing interference or loss of high frequency response in the picture. This is accomplished with special AFM heads and a new recording technique called *depth multiplexing*. Fig. 6.63 describes depth multiplexing. As the video head rotates, the A AFM (audio) head contacts the tape 120 degrees (5.53 milliseconds) before the A Y FM (video) head reaches the same spot. The AFM head first magnetizes the tape with an audio signal. The record current amplified to the AFM head is strong enough to create a magnetic field that penetrates the tape oxide coating. The A video head contacts the tape 5.53 milliseconds later but now the record current in the A video head is about half that applied to the AFM head. The magnetic field thus created is only strong enough to magnetize the top half of the tape oxide, so the deeper portion of the oxide coating retains the audio magnetic information while the upper portion is remagnetized to the video FM signal.

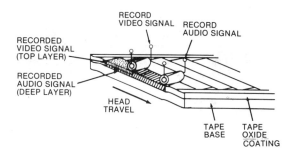

Fig. 6.63. Depth multiplexing.

In playback, AFM head A detects just the AFM signal. Azimuth record differences prevent the AFM head from recognizing the video head A information on the same track. The B heads operate in a similar manner. Depth multiplexing eliminates the need for VHS balanced mixers to create a separate pair of frequencies like that used in the Beta systems.

The placement of the heads on the drum is critical to proper operation. Some VHS manufacturers require replacement of the entire video head drum assembly including the upper head and lower drum motor when worn or defective heads are replaced. They claim this is necessary to ensure tight mechanical tolerances between the upper and lower drum assemblies. A possibility exists that if only the upper drum were replaced, the AFM and Y FM signals would not properly overlap. One disadvantage to this system may be the added cost associated with replacing the video heads.

Designing a pair of heads specifically for the AFM signals permits the use of simpler, less-sophisticated audio head switching circuitry. The playback signal is amplified by its own pre-amplifier. The audio head switching circuitry is similar to that for video head switching except audio head switching is done at a different time to compensate for the 120-degree advanced position of the AFM heads. The simplified AFM head switching circuitry does produce audible head switching noise. This slight popping noise is most noticeable when a steady 1 kHz tone is recorded and played back. Generally, the popping is not noticed during normal audio programs.

Regardless of the format, AFM provides oper-

Table 6.1. *Audio Specification Comparison*

MEASUREMENT	VCR TAPE SPEED	TYPICAL AFM SPECIFICATIONS	HIGH QUALITY AUDIO REEL-TO-REEL RECORDER	VCR LINEAR TRACK AUDIO SPECIFICATIONS
Frequency Response	SP	20—20,000 Hz	25—18,000 Hz	100—8,000 Hz
	EP (SLP)	20—20,000 Hz		100—5,000 Hz
Signal-to-Noise Ratio (SNR)	SP	60 dB	60 dB	42 dB
	EP (SLP)	60 dB	60 dB	40 dB
Total Harmonic Distortion	SP	0.5%	0.8%	6%
	EP (SLP)	0.5%		7%
Wow and Flutter	SP	.005%	0.04%	0.8%
	EP	0.005%	0.04%	*

*Normally not published (measurements of over 1% noted).

ational performance exceeded only by that of compact disc players. Table 6.1 is a performance comparison between AFM, linear track audio, and an expensive reel-to-reel audio tape deck. The AFM audio wow and flutter is practically nonexistent and its AFM frequency response exceeds that of reel-to-reel and linear track audio.

Frequency response is improved because writing speed is increased through the use of rotating heads. Wow and flutter are reduced because the slight tape speed variations that affect linear audio become insignificant. A reduction in tape speed from Beta 1 to Beta 2 (a reduction of one-half the linear speed) produces a very slight reduction in writing speed because the tape is traveling so slow in comparison with the rotating video heads that the speed ratio between the two is very high. Minor errors in tape speed which cause linear audio wow and flutter have little effect on AFM performance.

TUNERS

Most home VCRs have built-in tuners which allow them to operate completely independent of the TV. The incoming broadcast signal is connected to the VCR through a series of antenna connectors on the back. Using a standard rooftop antenna system, all local VHF (very high frequency) and UHF (ultra high frequency) channels can be received by attaching the incoming antenna signal to the appropriate VCR connector. The introduction of cable TV systems has caused confusion about proper hookup of the VCR. It's useful to understand how the television broadcast and community cable company channels are arranged in the broadcast frequency spectrum.

VHF, UHF, Mid- and Super-Band Channels

Broadcast TV audio and video signals are transmitted as radio frequency (RF) information. These frequencies vary depending upon the selected channel.

Fig. 6.64 shows the frequency spectrum for broadcast TV signals. Normal TV broadcast signals are divided into three blocks of frequencies known as VHF low band, VHF high band and UHF. VHF low band covers channels 2-6. VHF high band covers channels 7-13; and UHF covers channels 14-83.

Notice that a large frequency gap exists between channels 6 and 7. This space is used for FM radio broadcast, police, fire, aircraft, and radio signals. Cable TV companies also use frequencies in this gap to broadcast mid-band channels. Mid-band channels are placed between the VHF low and VHF high band channel frequencies.

Another large frequency gap exists between channel 13 and the first UHF station (channel 14). These frequencies are also used for radio communication. And cable TV companies use frequencies in this range just above channel 13 as super-band channels.

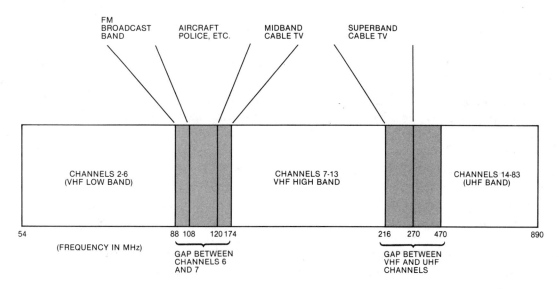

Fig. 6.64. Broadcast television frequency spectrum.

Confusion often exists between TV channels 14 through 83 which are broadcast in the UHF range and cable TV channels which may also be numbered from 14 up. Cable channel 14 is usually not the same as UHF channel 14.

Cable TV companies rarely use the UHF frequencies to send signals into the home. The signal strength of RF signals from the cable company decreases at a faster rate for higher channel frequencies than for lower channel frequencies because capacitance in the connecting cables causes an increase in resistive reactance as frequency increases. Signal strength losses are much more severe for every foot of cable in the UHF band than they are for the lower VHF bands. The higher the frequency of the broadcast channel, the greater the amount of signal strength lost per foot of cable. This is why cable companies often convert UHF signals to VHF or mid-band channels before sending them into the home.

Since cable TV systems don't use the UHF band TV channel numbers, they designate mid- and super-band frequencies using these same numbers. The cable company channel 14 may actually be a frequency found in the mid-band range; thus, it has no relationship to the original UHF broadcast channel.

Cable-Ready Tuners

Many TV receivers and VCRs have tuners that can receive only VHF low, VHF high and UHF channels. They cannot tune to the mid- and super-band frequencies. The popularity of cable TV systems was a catalyst for the introduction of special VCR tuner options called *cable-ready*. Unfortunately, the cable-ready designation has not been standardized in the video industry so a number of forms exist. Generally, a cable-ready tuner is designed to receive the mid- and super-band frequencies.

Cable companies frequently offer a cable conversion box that can receive the cable company mid- and super-band channels as well as the VHF low and VHF high band channels. This box allows incompatible TVs to receive the mid- and super-band frequencies. The cable box receives the mid- and super-band frequency stations and the other VHF channels and, then, converts these to a VHF low band frequency (typically channel 3 or 4) for use by the TV or VCR. The TV or VCR tuner receives these mid-band or super-band stations by tuning to channel 3 and selecting the desired station on the cable box.

Cable boxes also provide another useful function. Many of these boxes decode incoming special pay TV programs that are specially encoded at the cable company and are accessible for viewing by paying suscribers.

VCR cable-ready tuners will not decode these special pay channels, but they do offer the convenience of viewing all other nonscrambled cable TV channels without the use of the box. For this reason, a cable-ready tuner may be an option worth considering.

Hookup Configurations

VCRs connected to a normal rooftop antenna provide the opportunity to view one program while videotaping another by hooking up the VCR as shown in Fig. 6.65.

Incoming VHF signals are applied to the VHF antenna input terminal on the back of the VCR. UHF frequencies must be received by a UHF antenna and passed through associated cable to the UHF input on the VCR. The station that you want to record is selected on the tuner within the VCR. All channels arrive at the VHF antenna input on the back of the VCR. Suppose that you want to record channel 10 while watching channel 8. By pressing the TV/VCR button on the VCR and placing it in the TV position, channel 10 is received by the tuner within the VCR and recorded onto the tape. Another cable is hooked up from the VHF or RF output on the back of the VCR to the TV antenna terminals. This cable should be hooked up to the VHF input on the back of the TV. With the TV/VCR button on the VCR placed in the TV position, all channels available from the antenna at the VHF input to the VCR are also available at the output from the VCR and sent to the TV. Remember that channel 10 was selected for recording on the VCR. The TV tuner may be turned to any station. By selecting channel 8 on the TV, you can watch this station. By pressing the record button on the VCR you can record one station (channel 10) while watching another (channel 8).

If the TV/VCR button on the recorder is placed in the VCR position, only the station signal selected on the VCR tuner will be passed to the TV. In this case, the VCR is operating in the E-E mode described earlier in this chapter. The signal source selected on the VCR tuner is converted to channel 3 or 4 (same channel as the RF modulator) and is passed to the TV through the RF or VHF output connector on the back of the VCR. In this configuration, the only channel the TV can receive is the one coming from the VCR RF modulator (channel 3 or 4). In other words, when the TV/VCR button on the recorder is in the VCR position, only the program being received by the recorder can be viewed on the TV. When the button is in the TV position, all available channels are able to be selected using the TV tuner. In this mode, the VCR is independent of the TV.

With some limitations, this same feature may be used on cable TV systems. If the VCR doesn't have a cable-ready tuner, only channels 2 through 13 can be recorded. These tuners cannot receive the mid-band channels. If the TV/VCR button is in the TV position the VCR can record any station that the tuner can select while still allowing the TV to tune in on another station. The cable channels that the TV receives are also limited by the TV tuner. A TV without a cable-ready tuner can only tune in to the VHF low and VHF high bands from the cable system.

Fig. 6.66 shows a hookup configuration using a cable company conversion box with a VCR and TV that do not have cable-ready tuners.

All channels coming from the cable company are received by the cable box which converts each selected channel to a VHF low band frequency. In this example, channel 3 is the output of the cable box. The following steps show how you can record channel 10.

The cable box is tuned to channel 10. It converts the cable channel to a frequency corresponding to channel 3. Channel 10 audio and video is output as channel 3 from the cable box and fed

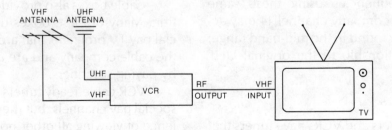

Fig. 6.65. Basic VCR hookup using conventional antenna.

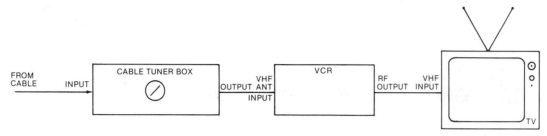

Fig. 6.66. Standard cable box-to-VCR hookup.

into the VCR. A cable is connected between the cable box and the the VHF input on the back of the VCR. To record the channel 10 program, the VCR must be tuned to channel 3 because the cable box has converted channel 10 to channel 3. In this configuration, only the VCR can receive channel 3 from the cable box. You can tune the cable box to channel 10, channel 8 or any other station but the cable box always outputs as channel 3.

In this configuration, you can only watch the station that is being recorded. Only one station is available at the input to the VCR and the input of the TV; this is the channel selected on the cable box.

With the system configured this way, you no longer have the option to record one station while viewing another. If the TV and VCR don't have cable-ready tuners, this configuration may be desirable, because the VCR and TV can receive and record the cable mid- and super-band frequencies. If a pay cable channel is coming into your home, the cable box will descramble these special channels and make them available for recording on the VCR.

Fig. 6.67 shows another option for interconnecting a cable TV system. In this configuration, the incoming cable signal is first connected to the VHF antenna input on the back of the VCR. The output from the VCR is applied to the input connector on the cable box. The output of the cable box connects directly to the VHF inputs on the TV. This will allow the recording of any cable channels which the VCR tuner can receive. Again, if the VCR doesn't have a cable-ready tuner, the channels are limited to the VHF low and high band stations. All stations including those in the mid- and super-band range are available at the input to the VCR, but the recorder cannot be tuned to receive the special channels.

If the TV/VCR button is placed in the TV position, all cable stations are simultaneously available at the input to the cable box.

This hookup is desirable if you wish to record a channel on a normal VHF station while watching a scrambled pay channel program on TV. This is accomplished by placing the VCR in the VHF station that you want to record, placing the TV/VCR select button in the TV position, tuning the cable box to the station you want to watch on TV and setting the TV to channel 3. Remember, the output from the cable box is always at the converted frequency (channel 3). If you have a cable-ready VCR, the hookup configuration in Fig. 6.67 is enhanced, and the VCR can easily record all cable channels except scrambled pay programs.

Fig. 6.68 describes a cable system configuration with increased options using two accessories. The first of these components is called a *splitter*; the other is called an *RF switch* or *A/B switch*. In this system arrangement, you can record all cable stations (including pay stations), watch those stations or a different station while you are recording a pay program.

The incoming signal from the cable is applied to a splitter that divides the signal equally and sends it in two directions. The first direction leads into the RF switch. If the RF switch is placed in the A position, the TV can select any station the tuner can receive. If a non-cable-ready tuner is in the TV, it will be limited to the VHF low and VHF high channels.

All cable channels are available at the input to the cable box. The cable box is selected to the station to be recorded, the selected channel is converted to channel 3, and passed to the VHF antenna input on the back of the VCR. In this configuration the VCR remains tuned to channel 3 regardless of the station to be recorded because the output from

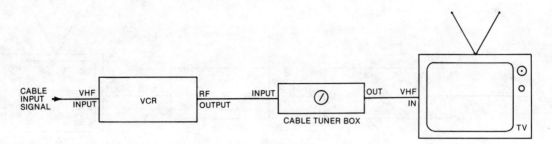

Fig. 6.67. Standard VCR-to-cable box hookup.

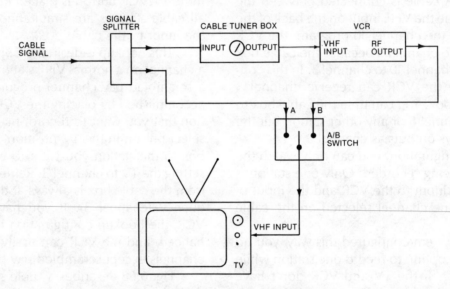

Fig. 6.68. Advanced cable hookup.

the cable box will always be broadcast on channel 3. The VCR can record any station selected on the cable box. The VHF or RF output jack at the rear of the VCR is connected to the B input on the RF switch. By placing the TV/VCR switch in the VCR position and the RF switch in position B you can watch the program being recorded by the VCR.

There are some important things to remember when using this configuration. When the RF switch is placed in the B position, the TV must be tuned to channel 3 (or channel 4 depending on the VCR RF modulator) to watch the program coming from the VCR. If the VCR is to perform an automatic timer record function, it must be commanded to tune in channel 3 at the time that the recording is to begin and the channel to be recorded must be previously selected on the cable box. If this is not done, the unit could record the wrong station or no station at all.

One disadvantage to this type of hookup is that multiple automatic record sequences using different channels can't be preprogrammed because the channel selector on the cable box must be tuned to the new channel. The VCR will automatically go on and off at its designated times but only the channel selected on the cable box will be recorded.

It is possible to use this configuration to view pay channels when the VCR is turned off. At that time, all signals available at its antenna input terminals are passed directly to the output VHF terminals and applied to the B input of the RF switch, so you can view any channel with the VCR de-energized using the cable converter box.

Many other configurations are possible by adding additional splitters and A/B switches, but picture quality deteriorates as cable splitters are added. Cable splitters divide signal strength in half for each dual output splitter. The number of splitters possible in a cable hookup configuration varies according to the strength of the signal coming in from the cable system. VCR accessories are available with a series of built-in A/B RF switches and signal splitters that provide flexibility and convenience.

An excellent reference for a wide variety of VCR hookup configurations can be found in the Howard W. Sams book *The Sams Hookup Book: Do-it-Yourself Connections for Your VCR*.

Hard of hearing consumers can record closed-caption TV programs using a caption decoder. Two types of screen caption formats are used in the United States—Line 21 and Teletext.

Line 21 broadcasts can be recorded directly from the antenna and then replayed through the caption decoder to generate the captions on the screen. You could also connect your antenna directly to the decoder and the decoder output to the VCR and then the TV set. This will cause the decoded caption images to be stored on the cassette tapes so they will appear on any VCR-TV system without further need for the decoder.

Teletext broadcasts can be recorded by placing the decoder between the antenna jack and the TV set. The VCR is configured to receive the signal from the TV and record the captions on tape.

Tuners and Audio-Video Demodulators

The tuner within a VCR is a specialized heterodyne frequency converter. All TV broadcast channel frequencies are simultaneously available at the input of the tuner. Fig. 6.69 is a block diagram of a VHF tuner. The channel frequencies are applied to an RF amplifier within the tuner. The RF amplifier is a frequency selective circuit that amplifies only the frequencies selected by the tuner. Each time a new channel is selected, the response characteristics of the amplifier are adjusted so the RF amplifier increases only the gain of the frequencies related to the selected channel. The amplified frequency is applied as one input to a heterodyne mixer circuit. The other input to the heterodyne mixer comes from a variable frequency oscillator that is adjusted to 45.75 MHz above the incoming selected channel frequency for the video carrier and 41.25 MHz above the audio carrier frequency for a selected channel.

Both sound and picture information are broadcast simultaneously using two separate carrier frequencies. The TV sound carrier frequency is always 4.5 MHz above the picture modulated frequency. The oscillator within the tuner is varied to the next correct frequency each time we select a different channel. Fine tuning adjustments make minor corrections to the oscillator in the tuner.

Picture and sound signals from the tuner are the sum and difference frequencies of the selected channel and the oscillator within the tuner. The tuner output is bandpass-filtered to pass the 41.25 MHz audio and 45.75 MHz video signals. All other frequencies are rejected. All stations tuned in by the tuner (whether UHF, VHF or mid-band frequencies) are converted to these new 41.25 MHz and 45.75 MHz frequencies.

The output from the tuner is processed, amplified and applied to two demodulators—one for audio and one for video. Audio and video signals are separated by frequency (remember they are broadcast 4.5 MHz apart) and applied to their respective demodulator circuits. The output of the audio and video demodulators are line level audio and 1.0 V peak-to-peak standard video. The audio and video signals are applied to their respective processing circuitry for recording onto the tape. The audio and video signals are simultaneously applied to the audio and video output jacks on the VCR as well as to the RF modulator where the signals are converted to channel 3 or 4 and available for output to the TV receiver. Many VCRs offer UHF and VHF tuners as well as the demodulator circuitry in a single assembly as shown in Fig. 6.70. Servicing the tuner demodulator sections is possible but replacement of the entire assembly is usually conducted by repair shops.

RF MODULATORS

An RF modulator is a tiny TV transmitter built into the VCR. Audio and video signals are applied to the RF modulator where they are frequency converted to a VHF low band channel. Two channels are usually provided on the modulator. A modulator output channel is selected by a switch accessible from the outside of the VCR. Two channels are provided so you can select a channel not in use in the local area. The RF modulator normally outputs on channels 3 and 4. If a channel is se-

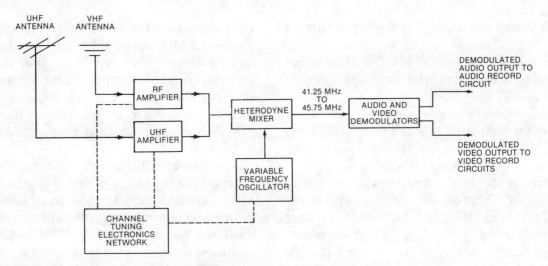

Fig. 6.69. Tuner block diagram.

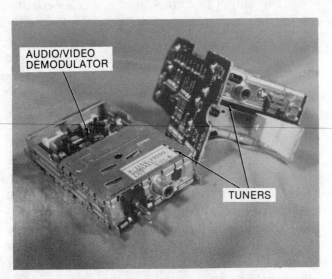

Fig. 6-70. Tuner and audio-video demodulation assemblies used in some VCRs.

lected which coincides with a broadcast station in the local area, interference may be present on the TV screen caused by the two stations beating against each other. If this occurs, the RF modulator should be changed to the alternate output channel.

RF modulators are carefully controlled by the FCC to prevent accidental interference to other nearby TV receivers. These low power transmitters have been carefully designed to transmit signals only within allowable frequencies. RF modulators are not considered serviceable items and are replaced as an entire assembly.

POWER SUPPLIES

VCRs operate almost entirely on DC voltage. The DC voltage is obtained by converting the AC voltage present at a wall socket. AC power from the wall electrical outlet is applied to the rectifier circuit within the power supply as shown in Fig. 6.71. Diodes are used to rectify the AC power by permitting only the positive-going excursions of the alternating current to pass. By configuring the diodes as a bridge rectifier circuit, it's possible to invert the negative-going portion of the AC sine wave and add it to the positive-going portion of the incoming signal. This effectively converts the 60 Hz full wave AC signal into a 120 Hz half wave AC signal as shown in the diagram.

The incoming AC power is applied to a transformer that lowers the effective AC voltage from 120 volts to a much lower AC voltage (normally less than 20 volts AC). This lower AC voltage is then converted to DC by the diode bridge rectifier.

The rectified AC 120 Hz signal is applied to a filter capacitor that charges to approximately the peak voltage at each of the rectified AC peaks. This capacitor tries to maintain this peak voltage causing a filtering action which removes all of the AC component from the input power. The AC signal is thus converted to an unregulated DC voltage level.

The raw DC voltage is applied to several voltage regulators that reduce the filtered (approximately 20 V) DC source to specific 12 V, 9 V, and 5

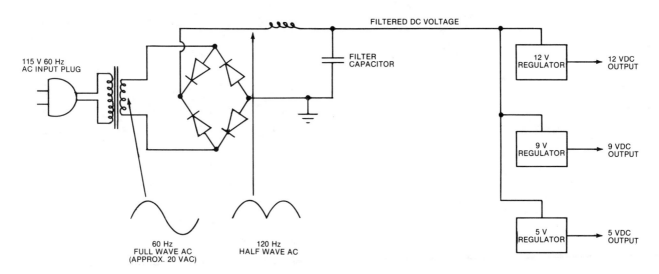

Fig. 6.71. Typical VCR power supply.

V levels. Some designs use the unregulated 20 volts to drive motors and relays which require higher operating current. 5 V supplies are typically used to drive the system control microprocessor and other integrated circuits. Twelve V and nine V supplies are used throughout the machine to operate various analog components.

Some power supplies have current limiting features in their design. If a short occurs in the VCR, the power supply is designed to limit the amount of current which can flow through the short circuit. In effect, these power supplies shut down until the short circuit has been corrected. This feature reduces the damage done to the power supply and other circuits by excessive current.

Power supplies are protected by fuses. The values of these fuses should never be changed because a higher amperage fuse will affect the safety rating of the circuit and no longer protect the circuitry. If a fuse continues to blow, even intermittently, there is a definite problem.

SUMMARY

So there you have it. In this chapter, the operation of Beta and VHS machines has been described to the circuit level. Block diagrams in the Appendix pull all of the subsystems described in this chapter into complete configurations for record and playback. Based on the detailed knowledge provided in this chapter, you should have a fairly strong understanding of how VCRs operate. Couple this with the chapters on troubleshooting and repair and you should be able to fix most VCR problems. At the least, you will know whether you can correct the problem by yourself, or if you will need a service center. And should you need to take your unit to a repair shop, you won't be easily mislead by fast-talking repair center operators.

CHAPTER 7

Advanced Troubleshooting

In earlier chapters you learned the basic techniques for troubleshooting most VCR failures. You found there are several steps to successful fault identification and correction. You learned how to recognize the various components of your VCR, and you discovered that a VCR fails in two ways, electronically and mechanically.

You learned that for technical or electronic repairs you use special tools to troubleshoot a circuit. This requires not only test equipment but some knowledge of electronics and VCR operating theory.

Here in Chapter 7 you will learn advanced troubleshooting techniques. You'll be introduced to the repair technician's "tools of the trade." Like other parts of this manual, Chapter 7 is full of "meat and potatoes" information to help you keep your VCR in peak operating condition.

Troubleshooting involves isolating a problem to a mechanical or electronic failure. If you conclude that the problem is not mechanical, you can use the techniques discussed in this chapter to test the electronic components in the failure area of your VCR.

TOOLS OF THE TRADE

When the problem can't be solved using flow charts and pictures, repair technicians reach for their tools. These include not only the tiny screwdrivers (tweakers), the diagonal cutters (dykes), and the soldering pencil, but also various measurement meters (ACVM, VOM, DVM, DMM), NTSC video generator, monitor/receiver, cross pulse monitor, frequency generator, variable power supply, frequency counter, oscilloscope and assorted gauges, jigs, and alignment tapes.

Meters

Electronic measurement equipment has improved a great deal over the years, markedly improving your ability to test and locate circuit troubles. Twenty years ago, a meter called a VOM (volt-ohm-millimeter) was used to measure the three parameters of an electric circuit—voltage, resistance, and current (Fig. 7.1).

Then came the VTVM (vacuum-tube-voltmeter). It wasn't long before electric circuits made room for electronic circuits, digital replaced analog gauges, and new meters appeared for

Fig. 7.1. A volt-ohm-millimeter (Simpson Electric Company).

troubleshooting. The DVM (digital-volt-meter) and DMM (digital-multimeter) quickly became the preferred measurement devices of digital technicians because they offered capabilities better suited for electronic circuit testing, including increased accuracy. These meters have characteristically high input impedances (resistances) so don't load down or draw down a digital circuit where the voltages and currents are far lower than those found in analog circuits.

Two changes affected the types of tools used in troubleshooting and repair. First, vacuum tubes were replaced by solid-state devices such as transistors and integrated circuits (ICs), or chips. Second, circuits themselves became smaller with more components packed compactly into less board area. You need only compare the early radios and televisions (standing four feet tall and weighing 40 pounds) with today's wrist radios and wrist televisions to recognize that electronic circuits are smaller, more complex, and more difficult to access by test probe.

AC Volt Meter

The AC volt meter (ACVM) is different from the previous meters described in that it is a special purpose volt meter. It can measure AC voltages ranging from only a few hertz to several hundred kilohertz. More expensive meters will measure into the low MHz range. Fig. 7.2 shows a DVM and an ACVM.

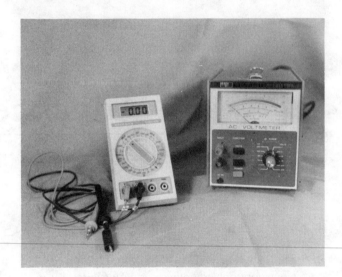

Fig. 7.2. The DVM and the ACVM test instruments.

An ACVM is used to measure audio voltages. It's required for the alignment of the audio circuitry. AC bias at the linear audio head is also aligned with this meter.

Audio AC bias adjustments require meters with an AC voltage sensitivity of one millivolt or less so the ACVM must be able to measure very small voltages at high frequencies.

NTSC Generator

NTSC stands for *National Television Systems Committee*. An NTSC video generator is a special purpose signal generator that produces precise video signals. Fig. 7.3 is a photograph of an NTSC generator.

The alignment of video circuitry requires stable video signals with horizontal and vertical sync, color burst, peak white levels, and accurate color phase and saturation levels. Proper alignment of the video circuitry is not possible using broadcast TV signals because they vary from one channel to another in strength, color saturation and peak-to-

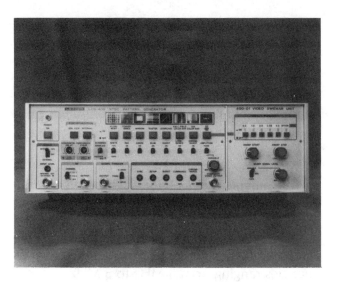

Fig. 7.3. NTSC video signal generator.

peak levels. If the VCR video circuitry were aligned according to the incoming broadcast TV signal, the accuracy of the alignment would depend completely upon the quality of the incoming signal and on the calibration of the VCR tuner and video demodulator circuit.

A second reason that broadcast signals can't be used to align video circuitry is that TV broadcast signals rarely contain 100 percent peak white video levels for more than an instant. Video alignment requires a one volt peak-to-peak video signal that can remain stable long enough for alignments to be made.

Monitor/Receiver

A monitor/receiver consists of a special purpose TV that can receive video signals through the standard cable or TV input connection and through special video input jacks that can connect to either a VCR or video generator. The monitor/receiver demodulates the station selected on the tuner and makes that modulated signal available at the video output jacks. These handy receivers are convenient because the demodulated TV signal can be applied directly to the video line input jack in the VCR, bypassing the VCR tuner and demodulator circuitry.

If a failure occurs in the VCR tuner or demodulator circuit, the ability of the unit to record and play back can be checked by injecting demodulated TV signals from the monitor/receiver directly into the VCR at the line input connector. This greatly simplifies troubleshooting.

The monitor/receiver can also receive and display a video signal directly from the VCR. The pure video line output signal from the VCR can be connected directly to the monitor/receiver video input. The video signal can then be displayed on the screen bypassing the RF modulator of the VCR and TV antenna switching electronics. If a failure is isolated to the VCR RF modulator, it can be located rapidly using the monitor/receiver.

A monitor/receiver also processes pure audio from the VCR audio line output. Demodulated audio from the station selected on the monitor/receiver tuner is also available at the audio output jacks and can be injected into the VCR audio line input. This is useful for testing the VCR audio circuitry. It allows you to bypass the audio demodulator in the VCR tuner circuitry.

The monitor/receiver can also verify and isolate audio record failures in the VCR tuner and audio demodulation circuits, or verify that the main portion of the audio record playback circuitry is performing properly.

Cross Pulse Monitor

A cross pulse monitor is a special type of video monitor used in analyzing VCR performance. This monitor is named for the effect produced on its screen. Horizontal and vertical synchronization pulses are displayed in such a way that they produce a figure resembling a cross as shown earlier in Fig. 5.19.

The cross pulse monitor provides information about the condition of the VCR by displaying the portions of a TV screen not normally seen. Delaying the horizontal and vertical synchronization pulses enables the observation of areas of the displayed broadcast TV signal that cannot be seen on a standard television receiver. Horizontal blanking, vertical blanking and synchronization pulses which are generated on the screen are made visible to the technician. The area of a TV picture in which video head switching occurs is also displayed on the cross pulse monitor.

Horizontal and vertical synchronization pulses and the time position of video head switching indicates the performance of the VCR during record and playback. A cross pulse monitor enables quick diagnosis of tape path error, servo error and RFI in the incoming record video signal. Worn video heads cause increased noise in this video picture at or near the video head switching point.

A common symptom of VCR malfunction is apparent servo circuit instability. During playback, the picture can jump up and down on the TV screen and the video tape can change speeds. This trouble can occur intermittently or on only certain channels. By viewing the E-to-E output signal from the VCR on a cross pulse monitor during record, the vertical and horizontal synchronization pulses can be analyzed for RFI interference. RFI mixed in the synchronization pulses (especially the vertical synchronization pulses) can cause servo instabilities. The servo circuit can become confused when RFI is present in the vertical synchronization pulse train and cause unstable picture recording.

The VCR uses the vertical sync pulse to reference the VCR head switching pulse and control track signal. The interaction between servo circuits and the vertical synchronization pulse were explained in Chapter 6. The cross pulse monitor is valuable to quickly identify servo difficulties.

Frequency Generators

Frequency generators align the audio circuits, the video record and playback circuits, and the video head pre-amplification circuit. The frequency generator used should have an oscillation range between 1 Hz and 5 MHz. Audio circuits need a steady tone during alignment of the record and playback amplification components. Audio frequencies range from 30 Hz to 20 kHz.

Audio sweep generators are popular because they can help a technician quickly align audio circuitry. A sweep generator automatically adjusts the audio frequency within a selected range. It can be adjusted to sweep between 30 Hz and 20 kHz several times per second. Once a sweep range has been selected, the sweeping is done automatically by the generator.

During alignment, a tone is recorded and then the amplifier electronics are adjusted to specifications listed in the VCR service manual. A second frequency is recorded, and more adjustments are made. Sometimes three or four frequencies must be recorded during alignment. These recorded signals are then played back and the playback amplifier is adjusted for equal output levels over all of these recorded frequencies. The sweep generator enhances alignment because all of the required frequencies are recorded simultaneously on the tape so the electronics can be aligned rapidly.

Video record amplifiers also require equalization alignment. The need for equalization was covered in Chapters 5 and 6. A high frequency oscillator is used to align the high frequency video record and playback amplification circuits. Typical video circuitry alignment procedures involve frequencies ranging from 3 MHz to 5 MHz.

It's not necessary to purchase a separate frequency generator for both audio and video. One frequency generator will do if the range is broad enough to cover both the audio and video frequency requirements.

Some frequency generators include sweep marker identification. Sweep marker identification uses an electronic marker superimposed on the output of the frequency generator to identify specific frequencies. Sweep marker generators can sweep a selected range and identify the position of certain frequencies within that range using markers. For example, setting the frequency generator to sweep between 1 MHz and 5 MHz sixty times per second, you could also cause the machine to identify the 2 MHz, 3 MHz, and 4 MHz positions by placing an electronic marker on the output signal when each of these frequencies are reached. By using an oscilloscope to view the sweeping output of the frequency generator, you can see tiny notches in the oscilloscope presentation at the 2, 3, and 4 MHz points.

Alignment procedures often require the adjustment of amplifiers for equal signal output levels at various frequencies. A frequency generator with sweep marker capability makes this alignment easy. The frequency generator is placed in the sweep mode and set to sweep the frequency range required in the alignment procedure. Markers are placed at the frequencies you are aligning. To perform the alignment, you adjust the controls as instructed for equal signal output at each of the marked frequencies.

Isolated Variable AC Supply

An isolated variable AC supply called a *Variac* is required for servicing some VCR problems. A Variac is an isolation transformer with one side connected through a rheostat. The isolation part of the transformer separates the VCR power supply from the incoming AC power supply.

Some VCRs have a hot chassis power supply that has one side of the AC line connected directly to the chassis. Test instruments (such as oscilloscopes) with ground probes that connect to the AC power line cannot be connected to machines with hot chassis power supplies without an isolation transformer because you could destroy both units. If you blindly connect test equipment to machines that have these type of power supplies, there is great potential for damage to the VCR, the test equipment and you.

The variable AC part of the Variac is useful in troubleshooting VCR shorts. The AC voltage applied to the VCR can be varied by the rheostat control on the Variac. When the Variac is placed in series between the wall plug and the VCR, the power applied to the VCR can be increased slowly from zero to the nominal 115 V. Use extreme caution and a Variac if you're going to troubleshoot the power supply section of your VCR.

Frequency Counter

A *frequency counter* counts individual pulses or sine wave cycles and displays the number of pulses or cycles that occur each second. Frequency is measured in hertz (cycles per second). Precise frequency is required for accurate adjustment of VCR color and servo circuits. An oscilloscope can also measure frequency, but not with the accuracy of the frequency counter.

Oscilloscope

The *oscilloscope* (frequently called *scope*) has been with us for years, although recent advances have added many capabilities to the instrument.

Simply put, an oscilloscope (Fig. 7.4) is an electronic display device that graphs signal voltage amplitude versus time or frequency on a CRT screen. A scope analyzes the quality and characteristic of an electronic signal by using a probe that touches a test point in a circuit. It also determines the voltage level of certain signals.

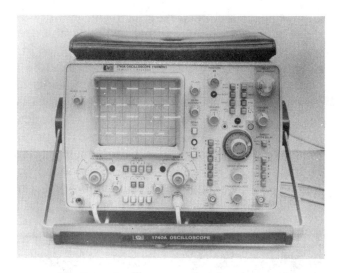

Fig. 7.4. Oscilloscope with delayed sweep. (Hewlett Packard)

Scopes come in all sizes, shapes, and capabilities. Prices vary between $500 and $20,000. Some scopes use a single test probe for displaying and analyzing a single trace signal. Others (dual trace) have two probes and display two different signals at the same time. On some oscilloscopes, as many as eight traces can be analyzed simultaneously. In fact, a recent electronics products magazine described the newest in oscilloscope technology, a seven-color digital scope from Test & Measurement Systems Company. Colors make it possible to rapidly compare signals at different locations in the circuitry. Some scopes even have built-in memories to let the machine store a signal of interest for future evaluation.

Besides sensitivity and trace display, a major distinguishing characteristic of oscilloscopes is that a wide range of frequencies can be observed on the CRT screen as frozen images. We call this range of frequencies *bandwidth*. Bandwidths vary between 5 MHz and 300 MHz, and price of the oscilloscope is proportional to frequency bandwidth.

Oscilloscopes can freeze an analog or varying signal and display this static waveform on a CRT screen covered with a measurement grid. While it is time-consuming to learn to use an oscilloscope, the analytical rewards are substantial. You can not only measure voltage amplitudes and frequencies of test signals, but also delay times, and signal rise and fall times. You can even locate the intermittent glitch.

Some oscilloscopes also have a variable delayed sweep function that expands the frequency being measured. It permits a low frequency signal to be displayed while expanding a small portion of that waveform for simultaneous viewing. Fig. 7.5 shows a delayed sweep scope display. The upper portion of the figure is a vertical field containing 262.5 horizontal lines. The bright portion of the vertical field is the section identified by the delay sweep function. The lower portion of the figure shows the expanded individual horizontal lines identified in the vertical field.

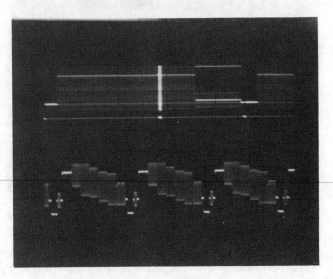

Fig. 7.5. A delayed sweep scope display.

For most VCR repairs you can usually get along fine with a dual trace, 25–30 MHz scope. Investment in an oscilloscope is a must if you intend to analyze and troubleshoot VCR component failures. If you intend to do only preventive maintenance, you'd be better off saving the money and letting a service center fix your machine when a scope is required

The nice thing about dual-trace, quad-trace, and even eight-trace capability is that it allows you to look at different signal paths or different signals simultaneously. For example, you could look at the input and output of a gate and actually see and be able to measure the delay time for the signal passing from input to output of the chip.

Torque Gauges and Tension Meters

Proper tape handling in the VCR prolongs the life of the video cassette tape. Torques applied by the reel tables to the tape during play, fast forward, and rewind must be controlled within specified limits to prevent tape damage. And the video tape must be wrapped around the rotating video head drum assembly with a specified amount of play or forward holdback tension.

If the torques are too low, tape may spill into the machine during operation. If the torques are too high, the tape can be stretched and worn by excessive friction.

Tension and torques are measured in different ways depending upon the machine. Torque measurement will be covered first.

Torques are normally not adjustable. An unacceptable torque measurement typically causes the replacement of parts. Low torques require replacement of rubber tires and belts. Excessive torques indicate the need to replace the torque limiting clutch. Brake torques which are out of specification normally require cleaning or replacement of the rubber brake shoes. Cleaning of the reel table assemblies is important for maintaining proper torques.

Torques are measured with a hand held torque gauge which fits over the reel tables as shown in Fig. 3.20 of Chapter 3. The machine is set to play with the torque gauge over the take-up reel table. As the reel table tries to rotate against the gauge spring, the meter needle deflects indicating the value of forward take-up torque.

Rewind and fast forward torques are measured by placing the torque gauge on the appropriate reel table and then engaging the machine into the mode to be checked. When you check rewind torque, place the gauge on the supply reel table (left hand reel table as you face the front of the machine). The right hand or take-up reel table is used to measure fast forward torque.

Brake torques are measured by setting the machine to STOP, putting the torque gauge on each of the reel tables and gently twisting the torque gauge counter-clockwise and then clockwise. The indications on the torque meter are noted for each of the reel tables and compared to specifications found in the service literature.

Tension is a factor that can be adjusted within your VCR. Tension is important for producing quality recordings and reproductions on a display screen.

Forward holdback tension is critical for correct playback. If the forward holdback tension is not within the allowable limits, the TV will display a flagging or tearing picture at the top of the screen. Vertical objects in the picture will bend (in some cases severely) at the top of the screen as you saw earlier in Fig. 4.11.

Flagging results when the horizontal automatic frequency control within the TV reacts to time base instabilities in the horizontal synchronization pulses from the video tape.

Holdback tension plays a significant role in reducing horizontal time base errors. Time base errors are discussed at length in Chapter 6. Video heads and tape guides will also wear faster if the forward holdback tension is excessive.

Using the tension gauge shown in Chapter 3, Fig. 3.19, set the VCR to play (use a T120 tape for VHS), and measure the tape at the point described in the VCR service manual. On most VHS machines, the forward holdback tension is measured near the point where the full erase track head contacts the tape.

The tension meter has three legs. The tape is placed between the three legs and the meter indicates the amount of holdback tension. Adjustment of holdback tension is usually made by setting the tension of a spring which holds a felt-padded brake band wrapped around the supply reel table as shown in Fig. 3.16 (Chapter 3).

Adjustment procedures vary according to the manufacturer and model of VCR that you own. Specific instructions are provided with the service literature.

Recently manufacturers introduced two cassette tapes like those shown in Fig. 7.6. Forward holdback tension and play torque are measured on one cassette, fast forward and rewind torques are measured on another. Separate gauges are attached to the hubs of the tape within these special cassettes.

Holdback tension and forward play torque are measured during normal play operation. Fast forward and rewind torques are checked by placing the machine in the mode to be tested. These cassette gauges are quite expensive but convenient, especially for those machines without the room for insertion of a hand-held gauge in the tape path. Such units require the cassette tape gauges.

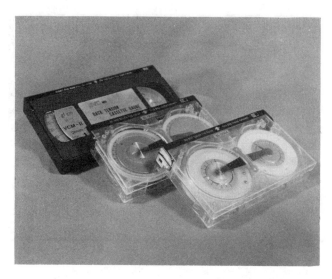

Fig. 7.6. Tapes are used to measure forward holdback tension, play torque, fast forward torque, and rewind torque.

Tension and torque measurements on Beta machines are usually made using tension and torque cassette tapes. Most VHS machines can be checked by using either hand-held gauges or the cassette tapes. In all machines, brake torque measurement requires the use of a hand-held torque gauge.

Tension and torques on machines with direct drive reel table motors are measured in the same way, but adjustments are made electronically.

Alignment Gauges and Jigs

When reel tables, tape guides, cassette tray assemblies, (and in some machines, video head drum assemblies) are removed, they must be aligned with accurate gauges and jigs after reinstallation. Some typical jigs and gauges are shown in Fig. 7.7 and Fig. 7.8.

Reel table heights are critical in avoiding tape edge damage. A reel table that is too high or too low will force the tape out of its normal traveling path and may cause the tape to be damaged at the top or bottom edge. Precise alignment of the cassette tape guides keeps the video tape in the correct path. Tolerances for cassette tape guides and reel table heights are measured within thousandths of an inch.

In many machines, the position of the cassette tray assembly is adjusted using a cassette tray alignment block. The block is placed into the cas-

Fig. 7.7. Typical jigs and gauges used in VHS alignment.

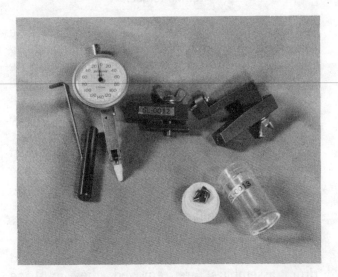

Fig. 7.8. Typical jigs and gauges used in Beta alignment.

Fig. 7.9. Use of the eccentricity gauge in video head drum assembly alignment.

drum assembly is aligned with an *eccentricity gauge*, a dial gauge with a soft plastic tip. The gauge is mounted to the chassis and the plastic tip is placed gently against the upper rotating head drum assembly as shown in Fig. 7.9.

The gauge must be placed over a part of the upper drum assembly where it won't contact the video heads as the drum rotates. The video head drum assembly is carefully rotated, by hand, and the deflection on the dial gauge is noted. Deflection indicates that the rotation is not uniform and that a slight wobble exists.

It's essential that the video head drum assembly rotates eccentrically. If the gauge indicates that this is not the case, the video head drum holdown screws are loosened slightly and the head drum assembly is repositioned. After the screws are retightened, the video head drum is again rotated carefully by hand and the movement on the dial gauge is monitored. This procedure is repeated until the dial gauge has minimum deflection over one full rotation of the head drum assembly.

The video head drum must rotate perfectly on its shaft or the video heads will not contact the tape evenly at all points. Poor head-to-tape contact results in snow on the screen.

Eccentricity adjustments are not made on all machines, but when these alignments are appropriate, they are done only after the heads have been replaced or removed.

sette tray after the tray has been removed from its position for accessing the reel tables, tires, pulleys, brakes, etc., contained beneath the tray assembly. With the alignment block inserted, the cassette tray is placed down into normal operating position. The block allows the tray assembly to be positioned for securing. Once in position, the screws holding the cassette tray assembly in place are tightened. This is simple, but important for proper cassette and tape path alignment.

On many Beta machines and some VHS machines, a replacement video head drum assembly must be aligned prior to use. The upper video head

Alignment Tapes

A factory alignment tape is an essential part of any major video repair. The alignment tape serves as both the mechanical and electrical standard for the VCR to ensure proper signal processing for both the audio and video signals and to confirm proper tape path alignment.

Factory alignment tapes are produced according to extremely tight standards set by the manufacturer. The VCRs which record alignment signals on these tapes are far more accurate than standard commercial machines. Strict control ensures that the test signals adhere to design specifications. Fig. 7.10 shows several alignment tape cassettes.

Fig. 7.10. The two front cassettes are used for alignment of Beta and VHS video cassette recorders.

Alignment tapes are expensive and should be used selectively, never in a machine whose mechanical condition is unknown. When you are testing a unit or making a coarse electronic or mechanical adjustment, use a good quality tape with a known good recording rather than the alignment tape. Once the machine is aligned close to its proper mechanical and electrical specification, use the alignment tape to fine tune the system to exact specifications.

The alignment tape contains several test signals including multifrequency tones for calibrating the audio head assembly and the audio playback electronics. High frequency audio signals are used to adjust the audio head azimuth position while low frequency audio signals are used to align the audio head height and playback electronics.

The alignment tape also includes special video signals for calibrating the playback video circuitry and video head pre-amplifier circuitry. Some tapes include special RF signals that are used only for tape path alignment. Alignment tapes that don't contain the special RF signals may still be used to align the tape path. In this case, any of the video test patterns on the tape can be used to adjust the tape path so it enables the electronics to produce the proper RF envelope.

Alignment tapes may also include audio FM signals to align VCRs with hi-fi stereo. Each service manual should contain a part number for ordering the alignment tape unique to that VCR. Most VCRs need different alignment tapes. Use only the alignment tape(s) specified for your machine.

The content of these tapes will vary depending on the video head core size of the target VCR and the particular audio option designed into the unit.

A specification sheet comes with the alignment tape to describe the calibration value of the particular tape. This calibration sheet is important and should be kept with the alignment tape. It describes the amount of inherent error in the tape speed and frequency of the recorded test tones. For example, if the specification indicates an alignment tape speed error of plus 0.1 percent, you should expect this error to be present in the playback speed of a machine under test that has been correctly calibrated. A factory alignment tape is a must for completely maintaining your own machine.

When performing a VCR alignment, the playback circuitry must be calibrated before any adjustments are made to the recording section. Playback alignment must be done first because the record alignment procedure involves playback of the recordings made during alignment. If the machine does not play back properly, playback checks of the recording alignment would be invalid. Once the playback circuitry is properly aligned, the record adjustments are then made.

COMPONENTS AND HOW THEY FAIL

While the use of troubleshooting equipment makes the analysis and isolation of VCR problems much easier, many failures can be found without expensive equipment. In fact, an understanding of how electronic components fail can make troubleshooting and repair relatively simple.

Other than for operator error, electronic failures generally occur in the circuits that are used or stressed the most. This includes the motor drive and power supply transistors and ICs. The microprocessor in the system control circuitry is highly reliable and doesn't fail very often. Most failures involve the other components which require soldering and are not as easy to replace.

Now that you understand the available tools, let's look at the kinds of components you'll be analyzing and how these components can fail.

Integrated Circuits—Chips

A chip or integrated circuit is fabricated from silicon with tiny particles of metal (impurities) imbedded in specific positions in the silicon. By positioning the metals in certain ways, the manufacturing process forms tiny transistors. Applying a voltage to specific places on the chip causes the device to invert a voltage level (+5 volts—logic 1, to 0 volts—logic 0 or vice versa), and enable all sorts of logic gates (AND, NAND, OR, NOR, etc.) to function. These chips can be made with silicon/metal junctions so tiny that today thousands of transistors can be placed on one chip. A memory chip the size of a fingernail can hold over 470,000 transistors.

The problem for chip manufacturers concerns how to apply voltages and get signals into and out of such a tiny chip. Very thin wires which are glued or bonded to tiny pads are used as inputs and outputs to the chip. The other end of each wire is bonded to a larger pad on a supporting material (the big part of what we call the integrated circuit as shown in Fig. 7.11). The supporting structure includes the pins we solder into the printed circuit boards.

These tiny silicon and metal chips are placed in environments that really put them under a lot of stress. They heat up when you use the VCR, cool down when you turn off the machine, then heat up again. This hot-cold-hot effect, *thermal stress*, affects those tiny strands of wire, or leads, going between the chip and the supporting structure (which includes the large pins that are inserted into the boards). Over time, thermal stress can cause the bonding of the wire lead to break away from the pad on the chip. This disconnect causes an input or output to become an *open* circuit, and chip replacement is required.

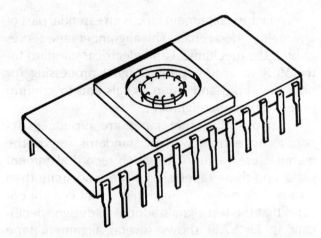

Fig. 7.11. The tiny leads from the chip to the pins of the chip package are clearly seen.

Another chip failure is caused by a phenomenon called *metal migration*. The chip can be compared to an ocean of atoms. Some tiny particles of metal float about in this sea, migrating in directions perpendicular to the electrical current flowing through the chip. Problems occur when these metal particles begin to collect in parts of the chip. If they concentrate in the middle of one of those microelectronic transistors, they cause the transistor to operate differently or not at all. If the resistance of these collected metals gets high enough, it causes the device to operate intermittently or to simply refuse to work. Since the failing transistor is part of a logic gate, the gate malfunctions and the output may become "stuck at 1" or "stuck at 0" regardless of the input signal. Theoretically, a wearout failure won't occur until after several hundred years of use. However, we shorten the life span of chips by placing them in high temperature, high voltage or power cycling environments.

Other problems occur outside the chip, between the chip leads and the support structure pin leads, the inputs and outputs of the device. These types of failures include inputs or outputs shorted to ground, pins shorted to the +5 V supply, pins shorted together, open pins, and connectors with intermittent defects. The most common trouble (assuming power is available) are opens or shorts to ground. Under normal use, chips finally fail with an input or output shorted to ground.

Capacitors

Understanding the way a standard capacitor is constructed will help you understand how these devices fail.

In Chapter 2, you discovered that there are several types of capacitors on the VCR boards. The capacitor is constructed of two separated plates. A voltage is placed across the plates and for an instant, current flows across the gap. But electrons immediately build up on one plate and cause the current flow to slow and then stop leaving the capacitor charged to some voltage potential. In addition to storing a charge, capacitors are used to filter unwanted signal spikes (sharp, quick peaks of voltage) to ground.

The *electrolytic capacitor* is constructed as shown in Fig. 7.12 below.

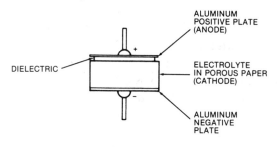

Fig. 7.12. The electrolytic capacitor.

Two aluminum foils or plates are separated by a layer of porous paper soaked with electrolyte solution, a conductive liquid. On one plate (the positive plate) a thin layer of aluminum oxide is deposited. This is called the *dielectric*. A capacitor has an anode (the positive plate), and a cathode (the electrolyte). Electrons build up on one plate causing it to become so negative that it prevents further current flow (remember that electrons have a negative charge).

Another type of capacitor is the *film* capacitor. It is constructed of alternating layers of aluminum foil and a plastic (usually polystyrene) insulation. The metal pieces of foil act as the plates and the plastic insulation acts as the dielectric between the plates. A film capacitor is shown in Fig. 7.13. Film capacitors are coated with epoxy and have tinned copper leads.

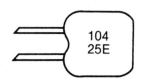

Fig. 7.13. The film capacitor.

Capacitors fail open or shorted depending on operating conditions and age. Electrolytic capacitors are especially susceptible to the aging process, one effect of which is the drying out of the electrolyte insulator. As the insulator dries out, the capacitance value decreases, and circuit performance decreases. Finally the capacitance value drops dramatically as the plates fold toward each other, and shorting occurs.

Another kind of failure occurs when some of the dielectric oxide dissolves into the moist electrolyte, causing the thickness of the dielectric to shrink. This deforming usually occurs when the electrolytic capacitor sits for a long time without applied voltage. In this case, the capacitance value increases but a large leakage of electrons occurs across the plates, making the capacitor useless.

The leads of a capacitor can physically detach from its plate causing an open in the circuit. Also, the plates can short together when a large area of one plate is stripped of its dielectric oxide layer by the application of too much voltage.

Despite the potential for several kinds of failure, capacitors in VCR circuitry seldom fail. However, capacitors can be *fried* by a momentary power surge through the power supply or damaged by excessive temperatures.

Resistors

These current-limiting, voltage-dropping devices are quite reliable and should function properly for the life of your VCR. However, the same factors that shorten the useful life of the chips also act to reduce the operational life of resistors. High temperatures, high voltage, and power cycling all affect the materials of which the resistors are made. These stresses cause breaks in the carbon, resistive paste, or resistive layers and produce an open conduction path in the circuit. Excessively high voltages can produce electrical current so large that it actually chars resistors to burnt ash. This is rare, especially in a circuit where the highest voltage seen is 12 volts (usually between 5 volts and 12 volts) and the currents are very tiny (milliamps).

Resistor failures are almost always associated with catastrophic malfunction of some other circuit component.

Diodes and Transistors

The diodes and transistors on the VCR printed circuit boards are made of solid material and act much alike. In fact, the transistor can be considered to be made up, in part, of two diodes.

Diodes are one-way valves for electric current, allowing current flow in only one direction. Diodes are usually made of either silicon or germanium. They are used in power supplies as rectifiers and in some circuits to maintain a constant voltage level. Other diodes are made of gallium arsenide and react by giving off light when biased in a certain way. These are called light emitting diodes or LEDs.

Transistors are used as amplifiers or electronic switches in various places in VCR circuitry. Diodes and transistors fail in the same ways and for the same reasons as chips.

USING TOOLS TO FIND FAILED COMPONENTS

Once the defect has been isolated to the failure of an electronic component, it's time to break out the schematic and block diagrams and turn on the test equipment. Because this guide was produced to cover all home VCRs, the block diagrams and troubleshooting flow charts found in this manual will help you understand and localize defects only to a particular circuit. The schematic diagrams and block diagrams found in VCR service manuals can help you isolate a failure to a specific defective part.

Troubleshooting the electronic circuits can involve tracing signals with a signal generator. These signals are injected into the audio or video circuitry, depending on the nature of the defect, and then traced to the failure point.

A TV monitor/receiver, an oscilloscope, a DVM, an NTSC video generator, an audio signal generator, a new, blank video tape, and a prerecorded test signal video tape are some of the basic tools you will need to find the problem. By comparing the symptoms with those found in the troubleshooting flow charts in this manual, you should be able to localize the defect to a particular circuit area. Much time will be wasted in troubleshooting if the problem is not localized to a specific circuit. Don't spend time looking for a problem in the audio circuit if your unit has no color. Locating failed components involves checking each of the input and output signals for proper DC voltages, peak-to-peak waveform levels and frequency (if it applies).

The following example will illustrate how to troubleshoot a brightness problem in the luminance processing circuitry. The first step is to decide if the defect occurs in playback, record or both. Assume that the defect occurs only in the record mode. You can verify that the VCR plays back properly by putting a video tape containing a good recorded signal in the machine and playing it back. If the signal plays back properly, you insert a blank tape and attempt to record. Place the VCR in the E-to-E record mode and attempt to monitor the signal coming from the VCR on your TV receiver.

In this example the program that you have tuned in on the VCR appears on the monitor/receiver, so you conclude that the VCR tuner, and audio and video demodulation circuitry are functioning correctly.

Next, connect the NTSC video generator to the video line input on the VCR and set the VCR to record the NTSC signal. Connect an oscilloscope by placing the clip lead of its signal probe at chassis ground. Take the measuring tip of the probe, place it on the video input line connector, and adjust the oscilloscope to receive a few horizontal lines of signal from the NTSC video generator at proper amplitude and frequency.

Check the block diagram in your service manual and compare that with the schematic diagram tracing the signal flow. Key test points will be called out on the block diagram or the schematic. Compare the waveforms observed on your oscilloscope with the waveforms shown in the schematic. Continue tracing the signal through the circuitry from input to output until an abnormality is found.

Eventually you will come to a point where the signal is no longer present or is deformed in amplitude or shape as compared to the schematic. This will be at or near the area of the defective component.

If waveform measurements do not isolate the defective component, use your voltmeter and check components in the area of the location of failure.

Measure and record the DC voltages around the ICs and transistors in the suspected area. Compare the measurements you made with those listed in the service manual. Individually check the components in the area for cracks and open connections along solder traces and connections.

Open connections prevent voltage levels from being transferred and prevent the affected circuit from being able to respond. If one input of a two-input NAND gate in the digital processing circuit has an open input as shown in Fig. 7.14, all but one of the four possible input combinations will be correct. With this type of failure, a VCR where only half of the inputs are good could operate correctly most of the time. The failure would be intermittent.

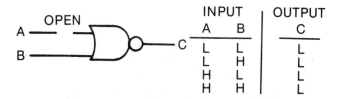

	INPUT		OUTPUT
	A	B	C
	L	L	GOOD
	L	H	BAD
	H	L	GOOD
	H	H	GOOD

Fig. 7.14. An open at the input to a NAND gate is only a problem in one of four logic-state cases.

As shown in Fig. 7.15, if the device being tested is a NOR logic gate, the output would be a logic "1" or "high" only when both inputs are at logic "0" or "low." Should one of the inputs become open, it would float to logic "1" and cause none of the input conditions to produce a logic "1" or "high" output. Thus, the output would be low all the time—just as though the output were shorted to ground.

If the chip has an open pin at its output, it can-

	INPUT		OUTPUT
	A	B	C
	L	L	L
	L	H	L
	H	L	L
	H	H	L

Fig. 7.15. An open at the input to a NOR gate will prevent the output from ever changing state or going HIGH.

not pass a logic "1" or "0" to the next gate. You can measure a voltage at the input to the next gate since it is providing the potential, a logic "1" or "high" level (something around +5 volts). The key here is that any time an input to a TTL gate opens (a condition we call *floating*), the gate will act as though a logic "1" were constantly applied to that input. The voltage on this floating input will drift between the high supply voltage of +5 volts and a level (about 1.5 volts) somewhere between a valid "high" and a valid "low." (A valid "high" is usually above +2.4 volts; a valid "low" is below +0.4 volts.)

A voltmeter reading of about +1.7 volts at the output pin of a gate on a chip is a clue that the output is floating open and the voltage is actually being provided by the next chip or following gate.

Since the VCR boards are flexible at certain points, replacing parts or depressing the boards without supporting them from beneath could cause a break to occur opening a trace on a circuit board. A hairline crack such as this is often difficult to find but looking at the board with a magnifying glass and a strong light (or a magnifying lamp) can sometimes reveal a suspected failure. A resistance test can be conducted with a VOM or VTVM by placing a probe at either side of the suspected bad trace as shown in Fig. 7.16, and observing whether a zero ohm reading is measured. Another way to ascertain if an open trace is present is to compare the voltage levels at either end of the trace.

Remember when you are testing for individual shorted or open parts in the VCR board circuitry that more than one part may use the same input or output lines to or from another component. When studying the circuitry, remember that the failure could be located at the other end of the board. One long trace from the end of the board to the part you are looking at may be shorted or open

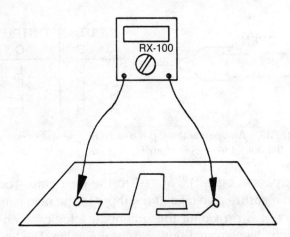

Fig 7 16 A trace can be tested for an open using an ohmmeter to test the electrical resistance from one end to the other.

at the distant end. Use of the schematics in the VCR service manual will help here.

OTHER TROUBLESHOOTING TECHNIQUES

There are some interesting tricks you can use to aid in finding chip failures.

Use Your Senses

Look, smell, and feel. Sometimes failed components become discolored or develop bubbles or charred spots. Blown devices can produce some distinctive smells, a ruptured electrolytic capacitor, for example. Finally, shorted chips can get really hot. By using a "calibrated finger," you can pick out the hot spots on your board.

Heat It, Cool It

Heating and then cooling is a fast technique for locating the cause of intermittent failures. Frequently, as an aging device warms up under normal operation, it becomes marginal and then intermittently quits working. If you heat the energized area where a suspected bad chip is located until the intermittent failures begin, then methodically cool each device with a short blast of canned coolant spray, you can quickly cause a marginally defective chip to function again. By alternately heating, cooling, heating, and cooling, you can pinpoint the trouble in short order.

You can heat the area with a hair dryer or a focused warm air blower designed for electronic testing. Be careful using this technique, because the thermal stress you place on the chips being tested can shorten the life of good components. A one- or two-second spray of freeze coolant is all you should ever need to get a heat-sensitive component working again.

Most coolant sprays come with a focus applicator tube. Use it to pinpoint the spray. And avoid spraying electrolytic capacitors, because the spray soaks into the cap and destroys the electrolyte in some aluminum capacitors. Also be careful not to spray your own skin. You could get a severe frost burn.

The Easter Egg Approach

Quite often we can quickly locate a fault to a couple of parts but need further testing to determine which one is the culprit.

When time is of essence, take an *Easter Egg* approach. Just as a youngster used to pick up and examine Easter eggs one at a time to see if his/her name was marked on it, you can try replacing the chips one at a time to determine whether the chip replaced was causing the problem. You have a fifty-fifty chance of selecting the right part the first time. If that doesn't work, replace the other components.

If the parts involved are inexpensive, why not replace them both? For thirty cents more, go ahead and splurge. If the problem's gone, but you're still curious, you can always go back later and test each part individually.

A typical problem occurs in oscillator circuits where a failure can be isolated to one of several capacitors. A fast way to restore proper signal is to replace all of the suspected capacitors.

Testing Capacitors

How do you check out a capacitor that you believe has failed? If the device has shorted, resulting in severe leakage of current, you can spot this easily by placing an ohmmeter across the capacitor and reading the resistance. At first you'll notice a low reading, because the capacitor acts as a short until it charges; but then, if the capacitor is working properly, it will charge, and the resistance will rise to a nominally high value. If the device is shorted, the initial low resistance reading continues and the capacitor won't charge.

Should the component be open, you'll not see the instantaneous short at time T0, the moment charge starts to build. An open circuit has infinite resistance. An in-circuit capacitance tester is helpful here.

Total failure as a short or open is pretty easy to find. But how about the device whose leakage depends on temperature or whose dielectric has weakened, changing the capacitance value? To test this capacitor requires a different level of analysis.

Capacitance Measuring

If you have an ohmmeter whose face has the number 10 in the middle of the scale, you can easily use it to approximate the capacitance of a device. Use the time constant formula $T=RC$, where T equals the time in seconds for a capacitor to charge to 63.2 percent of supply voltage, R equals the resistance in ohms, and C equals the capacitance in farads. Using a 22 μF (.000022 farad or 22 microfarad) capacitor and a 1 M ohm (1,000,000 ohm or 1 mega ohm) resistor, the charge time for one time constant is .000022 $\times$ 1,000,000 = 22 seconds.

Transposing the formula to read:

$$C = T/R,$$

we can determine the value of capacitance by knowing the resistance and counting the seconds required for the charge to cause the ohmmeter needle to reach 63.2 percent of full scale (infinite resistance). This point is at about 17 on the meter.

To do this procedure, disconnect one end of a capacitor from the circuit, turn on your meter, and let it warm up for a minute. Zero adjust the ohms scale reading. Then estimate the ohms scale multiplier needed to let the capacitor charge in some acceptable time period. For microfarad capacitors use the $\times$ 100K scale because this will let the capacitor charge in less than a minute. The 17 on the scale represents 1.7 megohms on the $\times$ 100K scale.

Short a low-ohm-value resistor across the two capacitor leads for several seconds to thoroughly drain off any charge. Then connect the ground lead from the meter to the negative side of the capacitor (either side if the capacitor is not an electrolytic), and touch the positive meter probe to the other side of the capacitor. Using a stop watch to count seconds and tenths of a second, watch the face of the ohmmeter as the capacitor charges and the resistance needle moves up. When the needle gets to 17 on the scale, stop the clock and read the time. This will give you the capacitance value in microfarads.

This technique will give you a close enough approximation of the capacitance value to determine if the device is good or should be replaced.

Some capacitor failures only occur when a voltage of about 50 percent of their rated working voltage is applied because they might fail only under operational voltage and temperature loads. Yet when they're removed from the circuit and tested as described, they appear to be good. If they are resoldered back in the circuit, voltage measurements or scope presentations of signals in the circuit again point to the capacitors as likely problem candidates. If you think a capacitor is defective and it tests good out of circuit, substitute it with a new capacitor to determine the condition of the original part. Do this for each suspect capacitor.

Replacing Capacitors

Always try to use the same type and value capacitor as the one being replaced. Keep the leads as short as possible and solder the capacitor into the solder connector holes with the proper iron. The solder process should not exceed 1.5 seconds per lead, otherwise, heat damage to the component may result.

A good technique to use is to tin the capacitor leads just before poking them through the circuit board holes. This speeds the solder bond process.

Testing Diodes

If you have a digital multimeter (DMM) with a diode test capability, you can quickly determine whether a suspected diode is bad or good. Placing the meter on the ohmmeter setting and positioning the probes across the diode causes the meter to apply a small amount of current through the diode if the diode is forward biased. The voltage drop across a diode is normally 0.2–0.3 volt for germanium diodes and 0.6–0.7 volt for silicon diodes. Reversing the leads should result in no current flow, so a higher resistance reading should be observed. A low resistance reading when the diode is biased in either direction indicates that the device is leaking or shorted. A high resistance reading in both directions indicates the diode bond has opened. In either case, replace the diode immediately.

Diodes can also be tested in-circuit using the ohmmeter to check the resistance across a diode in both directions. With one polarity of the meter probes, you should get a reading that is several hundred ohms different from that obtained when the probes are reversed. For example, in the forward-biased direction, you could read 50–80 ohms; in the reverse biased direction, 300K ohms. This difference in readings is called DE for *diode effect* and is useful for evaluating transistors. When diode readings in both directions show low resistance, you can be sure the leaky short is present.

Testing Transistors

It's no fun to desolder a transistor, test it for failure, find it good, and solder it or a new device back into the circuit board.

Fortunately, there is a way to determine the quality of silicon transistors without removing them from the circuit. In 90 percent of the tests, this procedure will accurately determine whether a device is bad.

A transistor acts like a configuration of diodes as shown in Fig. 7.17. PNP and NPN transistors have opposite-facing diodes. The transistor functions by biasing certain pins and applying a signal to one of the leads (usually base) while taking an output off the collector or emitter.

The following tests apply to both PNP and NPN transistors. If an ohmmeter is placed between the collector (C) and emitter (E) as shown in Fig. 7.18, it effectively bridges a two-diode combination in which the diodes are opposing. You should get a high resistance reading with the leads applied both ways. (It's possible to wire the transistor in a circuit which makes the transistor collector-emitter junction act like a single diode. In this case you could get a DE. Both results are O.K.)

Typical C-E resistance readings for germanium transistors are as follows:

Forward biased = 80 ohms
Reverse biased = 8000 ohms (8K)

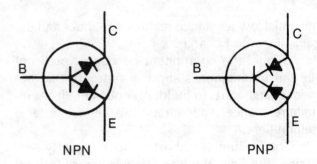

Fig. 7.17. A transistor acts like a pair of diodes.

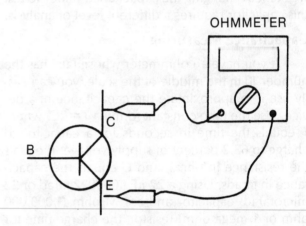

Fig. 7.18. A transistor can be tested using an ohmmeter placed across the collector-to-emitter junction.

For silicon transistors you might read:

Forward biased = 22 M ohms
Reverse biased = 190 M ohms

The high/low ratio is evident and is about the same for both types.

Place the probes across the collector-to-base junction leads. Reverse the probes. You should observe a low reading in one case and a high reading with the test probe leads reversed (the diode effect).

Try the same technique on the base-to-emitter (B-E) junction lead (Fig. 7.19). Look for the DE. If the DE is not present in all the preceding steps, you can be certain the transistor is bad and needs to be replaced.

Another way to evaluate a transistor is to measure the bias voltage from base to emitter on an energized circuit. Confirm the correct supply voltage first; power supply problems have been

Advanced Troubleshooting 213

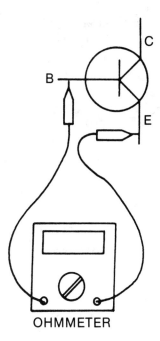

Fig. 7.19. Check the base-to-emitter junction for diode effect (DE).

known to trick troubleshooters into thinking a certain component has failed.

The B-E forward bias for silicon transistors should be between 0.6 and 0.7 volt DC. If the reading is below 0.5 volt, confirm that the transistor is a silicon type from the parts list and, if it is, replace it. The diode junction is leaking too much current. If the reading is almost a volt, the junction is probably open and the device should be replaced.

B-E Voltage (Forward biased)	Action
0.5 V	Replace
0.6–0.7 V	Good, keep
0.9	Replace

Although in some isolated cases some other failure could cause the low reading, the most common cause of low bias voltage is failure in the transistor itself.

If the previous tests are inconclusive, there is something else you can try. Measure the voltage across the collector-to-emitter (C-E) junction. If the reading is the same as the source supply and you notice on the schematic that there's plenty of resistance in the collector/base circuit, the junction is probably open. Replace the device.

If your reading is close to 0 volts, take a small length of wire and short the base to the emitter, removing all the transistor bias. The C-E meter reading should rise instantly. If it doesn't, the transistor is shorting internally and should be replaced. If C-E voltage does rise, it suggests a failure in the bias circuitry, perhaps a leaky coupling capacitor.

Removing Solder

One way to remove residual solder is to use a *solder sucker*—a hand held vacuum pump with a spring-driven plunger—to pull the hot, melted solder off a connector (Fig. 7.20). Heat the old solder until it melts, place the spring-propelled vacuum pump in the hot solder, quickly remove the soldering iron while you release the vacuum pump's spring, and suck the solder up into a storage chamber in the pump.

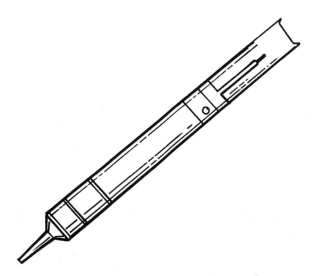

Fig. 7.20. The spring-driven plunger in the solder vacuum pump is used to pull hot solder off a connection.

This technique works fine until you try to use it around CMOS chips. Some vacuum pumps produce static electricity, and by now you know what that can do to an MOS or CMOS chip.

A safer way to remove solder is to touch the solder with the end of a strip of braided copper. Then heat the braid just a short distance from the solder (Fig. 7.21). The copper braid heats quickly, transferring the heat to the solder, which melts and is drawn into the braid by capillary action. Then, cut off the solder-soaked part of the braid and throw it away.

Fig. 7.21. Use of a solder wick is another way to remove solder from a connection.

Removing components from double-sided circuit boards is difficult without the use of a special desoldering iron like the one shown in Fig. 7.22. This iron has a hollow point and uses a motorized vacuum pump to heat up and suck the solder into a reservoir. This desoldering tool makes components that have been soldered on both sides of the board much easier to remove. It's also safe for static sensitive parts.

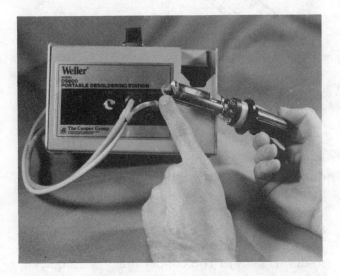

Fig. 7.22. A hollow-point soldering iron with a motorized vacuum pump to heat up and suck solder into a reservoir.

If any solder remains in the circuit-board hole, heat the solder and push a toothpick into the hole as the solder cools. The toothpick will keep the hole open so you can easily insert another wire lead for resoldering.

Be careful not to overheat the board during the solder-removal process. Excessive heat can cause part of the circuitry to come away from the board. It can also damage good components nearby.

If you remove the solder from a component and a lead is still stuck on some residual solder, pinch the lead with a pair of needle-nose pliers as you gently wiggle it to break it loose from the solder bond.

The pins of most chips on VCR printed circuit boards are bonded to the circuit board by a process called wave soldering. Wave soldering produces an exceptionally good bond without the added manufacturing expense of a socket. This process helps keep the fabrication costs down, but it makes it more difficult for you to replace the chip.

One effective way to remove wave-soldered chips is to cut the chip leads or pins on the component side and remove the bad chip. Then remove the pieces of pin sticking through the board using a soldering iron and solder braid or a vacuum pump.

Some special tools are available to help you remove soldered components. Fig. 7.23 shows a desoldering tip that fits over all the leads of a chip or dual-in-line package (DIP) device.

Fig. 7.24 is a photograph of a spring-loaded DIP extractor tool. By attaching this device to the surface mounted chip on the printed circuit board and then applying the DIP tip shown in Fig. 7.23 to the soldered connections on the opposite side of the board, you can easily remove the chip. When you press the load button downward and engage the clips, you will cause the extractor to place an upward spring pressure on the chip. When the solder on the reverse side melts enough, the chip will pop up and off the board.

Fig. 7.25 is a photograph of a *flat pac* IC. This low-profile device is mounted directly onto the circuit board and removal requires a special desoldering tip to fit over the IC and heat all leads at the same time. With different size ICs, a variety of desoldering tips may be required.

Advanced Troubleshooting 215

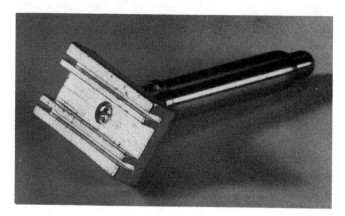

Fig. 7.23. A desoldering tip for removing chips that are soldered to the circuit board.

Fig. 7.25. A flat pack integrated circuit.

Fig. 7.24. A spring-loaded dual-in-line extractor tool.

SOLDERING TIPS

No pun intended. Hand soldering is the most misunderstood and most often abused function in electronics repair. Not only do many people use poor soldering techniques, but they also use the wrong soldering irons.

Solder isn't simply an adhesive making two metals stick together. It actually melts and combines with the metals to form a consistent electrical as well as mechanical connection. Time and temperature are critical in this process. The typical hand solder job can be accomplished in 1.5 seconds or less if the soldering iron and tip are properly selected and then properly maintained.

The nominal solder melting temperature is 361 degrees F. Metal combination between the solder and the metals being joined occurs at temperatures between 500 and 600 degrees F.

Most soldering jobs join the metals copper and tin, but both of these metals are easily oxidized. Poor or no solder connections are made if the surfaces to be connected are covered by contaminants such as oils, dirt, or even smog, so be sure to use solder with a good cleaning flux. The flux prepares the surfaces for best solder metalization. The flux melts first and flows over the metal surfaces removing oxidation and other contaminants. Then the metal heats so that the solder melts and flows, producing a good, shallow bond.

The key to successful soldering is in the soldering iron tip. Most people selecting their first soldering iron buy a low wattage iron, but this is a mistake. Instead, pick an iron whose tip operating temperature is suited for the circuit board you're to repair. If the tip temperature is too low, the tip sticks to the surface being soldered. If it's too high, it damages the board surface. The ideal working temperature for soldering on a VCR circuit board is between 600 and 700 degrees F.

The soldering iron tip is used to transfer the heat generated in the iron out to the soldering surface. The iron should heat the tip quickly, and the tip should be as large as possible yet slightly smaller than any soldering pad on the board.

Tips are made of copper. Copper conducts

heat quickly, but it dissolves in contact with tin. Solder is made of tin and lead. To keep the tin from destroying the copper tip, manufacturers plate a thin layer of iron over the soldering tip. The hot iron (now you know where the term *iron* came from) still melts the solder, but now the tip lasts longer. The iron melts above 820 degrees F, so if the heat produced by the iron stays below 700 degrees F, the solder melts but not the iron plating.

The disadvantages of the iron plating are that it doesn't conduct heat as well as copper, and it oxidizes rapidly. To counteract this, you can melt a thin coat of solder over the tip. This is called *tinning*. This solder layer helps the soldering iron heat quickly and also prevents oxidation.

The tip of an old soldering iron is usually black or dirty-brown with oxidation. And it doesn't conduct heat very well. These "burned-out" tips can be cleaned with fine emery cloth, retinned, and then used.

Wiping the hot tip with a wet sponge just before returning the iron to its holder is a mistake. This removes the protective coating, exposing the tip surface to atmospheric oxidation. It's much better to add some fresh solder to the tip instead. Keep your iron well tinned.

Fig. 7.26 shows the proper way to solder a plug or connector lead. Place the tip of the iron on one side of the lead and the solder on the other side.

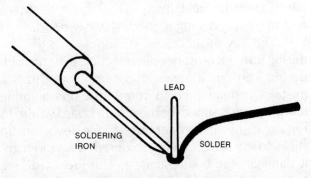

Fig. 7.26. Place the soldering iron on the opposite side of the lead from the solder.

As the solder pad heats, the tin-lead solder melts and flows evenly over the wire and the pad. Keep the solder shallow and relatively even. When you think your soldering job is complete, carefully inspect your work. Sometimes, if you aren't careful, you can put too much solder on the joint, so that there's not enough solder on the top or bottom of the connection. It's also possible to get internal voids or hollow places inside the solder joint. Large solder balls or mounds invite *cold solder joints* where only partial contact is made. Fig. 7.27 shows some examples of inadequate soldering. These kinds of solder joints can be a source for intermittent failure.

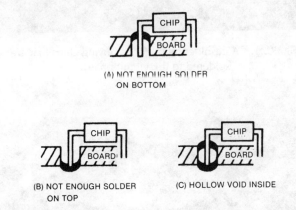

Fig. 7.27. Examples of inadequate soldering on a printed circuit board.

Good soldering takes patience, knowledge, and the right tool—a grounded, temperature-controlled soldering iron with a tip temperature maintained in the 500–600 degree F range for optimum soldering.

Before You Solder It In

A useful thing to do before you solder in a replacement part is to test the device in the circuit. Simply insert the chip or other device into the solder holes and wedge each lead in its hole with a toothpick. Then energize the circuit and test. After proper function is assured, remove the toothpicks and solder the component into the board.

Circuit Board Repair

Repairing damaged circuit boards is a lucrative business, and several companies have developed around this activity. For some board failures, you can repair your own circuitry and save some money.

Before soldering in new components, check over the board for any broken traces or pads lifting off the board. If a trace is open and is starting to lift away from the board surface, jumper across the broken spot from one component solder pad to another pad. Use solid #18 or #20 wire tinned at both ends before soldering.

If a pad or trace lifts free, replace it with an adhesive-backed pad or trace overlapping the damaged area. Scrape the coating off the pad or both ends of the trace so the new pad or trace can be soldered firmly to the existing pad or trace. Remove all excess solder and redrill any lead hole that has become covered or plugged with residual solder.

Recommended Troubleshooting and Repair Equipment

If you're planning to tackle failures that usually require service center support, you can minimize your investment costs and yet optimize your chance of success by carefully selecting your equipment and tools.

First, get a set of good metric screwdrivers—both Phillips and flat head. VCR screws are metric and require special metric screwdrivers. Get a wide selection of sizes, from the tiny *tweakers* to an 8-inch flat head. You might also find a set of jeweler's screwdrivers quite helpful. You will need standard metric nut drivers and metric hex-nut drivers of assorted sizes.

Then get several sizes of long-nose or needle-nose pliers. Get several sizes of diagonal cutters or *dykes* for cutting wire and pins. A good, low wattage, well-grounded soldering iron whose tip temperature is automatically controlled is a must if you intend to replace components. A simple 3½-digit DVM or DMM is useful for test measurements.

If you expect to do any serious troubleshooting, get a 30–50 MHz oscilloscope with dual trace, delayed sweep and a time-base range of 200 nanoseconds to a half second. Select a scope with a vertical sensitivity of 10 millivolts per division or better.

Fig. 7.28 is a photograph of typical hand tools used in the maintenance and repair of Beta VCRs.

Fig. 7.29 is a photograph of typical hand tools used in the maintenance and repair of VHS machines.

Table 7.1 shows an approximate price list for troubleshooting and repair equipment.

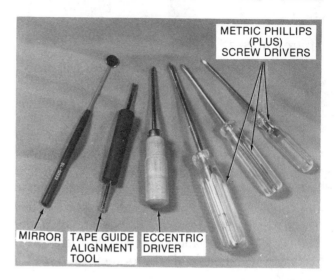

Fig. 7.28. Common Beta hand tools.

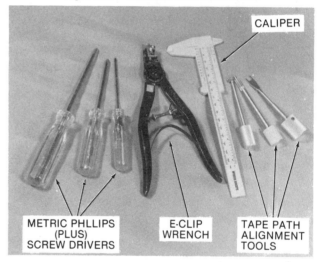

Fig. 7.29. Standard VHS hand tools.

Table 7.1. *Approximate Prices for Troubleshooting and Repair Equipment*

Metric Screwdrivers - 12		$35.00
Pliers		15.00
	4½" short nose	
	5¾" long nose	
Diagonal cutters		10.00
	4½" flush	
	4½" midget	
DMM (3½ digit)		80.00
Tension meter		400.00
Torque gauge		350.00
Desoldering station		300.00
Soldering iron		125.00
Oscilloscope		1800.00
NTSC generator		1000.00
Sweep function oscillator (1 HZ–5 MHz)		750.00
AC Voltmeter		350.00
Alignment tape		300.00

Spare Parts

Because of the cost involved you will probably want to maintain a minimal stock of repair parts; yet you want to be able to fix your machine quickly when it breaks down.

The most practical backup would limit your spare parts to tape end sensor lamps, belts, rubber tires and torque limiter assemblies. Your VCR is much more likely to experience a mechanical failure before an electronic malfunction occurs. These spares represent an investment of under $50.00. Several manufacturers sell spare parts packages with belts and tires for their various VCR models.

SUMMARY

There are three possible ways to optimize your VCR's operational life:

1. Buy a highly reliable unit with a good track record of performance.
2. Do preventive maintenance regularly.
3. Become a knowledgeable repair technician yourself.

Armed with the knowledge in this manual, you'll be able to spot downright poor troubleshooting like the "tech" rubbing up and down on the video head using low-grade alcohol to clean it, the repair person wiping his or her soldering iron on a wet sponge just before putting it in its holder. These are mistakes of poorly trained (or poorly motivated) people working on someone else's machine. You'll also be able to recognize the sharp, highly trained technician who uses the right tools and the right procedures to troubleshoot and repair in minimum time. Then you'll smile to yourself, knowing that you were smart enough to buy this book and do your own repair the right way.

Glossary

a Type Alpha wrap.

ACC Automatic Color Control. Used to maintain constant color signal levels.

ACK Automatic Color Killer.

Address Track A longitudinal track running through the video tracks (usually the vertical blanking interval) which is used for recording cue or time code information.

Adjacent Track The video track to the immediate right or left of a selected track.

AFC Automatic Frequency Control. Used to phase-lock the color circuitry to the recording or playback color signal.

AFM Audio Frequency Modulation.

AGC Automatic Gain Control. Used to maintain a fixed output signal level in luminence and audio circuits.

Alignment Tape A prerecorded tape that serves as a reference for electrical and mechanical performance evaluation of a VCR.

Alpha Wrap The name given to the complete overlapping turn of tape around the video drum as it forms the Greek letter, alpha. This type wrap allows a single head to be used for recording.

Amplifier Circuitry which boosts or increases the strength of an input signal.

AND Gate A logic gate in which all inputs must be High (Logic 1) to produce a logic High output.

APC Automatic Phase Control. Used to phase-lock color circuits to the recording or playback signal to achieve color signal stability.

APL Automatic Program Location. The use of electronic markings on video tapes to enable the locating of the beginning or end of a recording by electronic means.

APL Average Picture Level. A measure of the average luminance (brightness) level of a screen image.

Assemble Edit Constructing a video recording by adding scenes in sequence.

Audio-dubbing Adding a new sound track after the original recording has been made. The sound track is erased and rerecorded. The video and sync tracks are unaffected.

Audio Track The longitudinal portion of the tape on which audio information is recorded using a stationary head.

Automatic Rewind Electronically rewinding the tape to the beginning of a cassette.

Azimuth The left or right tilt of the gap of a recording head that enables extended playing and recording. The angle measured CW from a perpendicular formed by the head gap and the direction of the track across the tape.

Azimuth Loss The loss in picture quality caused by the oblique angle relationship of the recording and playback heads.

Balanced Modulator A circuit that produces an output comprised of the sum or difference between two input signals. The output signal will contain any special characteristics in one of the input signals. No output is produced when the chrominance signals are zero.

Bandwidth A range of frequencies over which an output remains uniform; measured between the 3 dB power point (point where power drops to half) or 6 dB voltage point.

Betamax A video recording format developed by Sony.

B-H Magnetization Curve A graph describing the magnetic effect on flux density (B) of an applied magnetic field of intensity (H).

Blanking The retrace time when the TV scanning beam shifts from the lower right to the upper left of the screen while the electron beam is de-energized.

Blanking Gap A vertical interval between two separate video frame signals.

Braking, Magnetic A servo control using EMF to slow down or speed up a drive shaft.

Brightness The total energy in the Y luminence signal.

Burst An 8 to 10 cycle occurrence of the chrominance 3.58 MHz subcarrier signal that appears after the horizontal sync and centers on the blanking portion of the video waveform. Used as the reference for phase comparison during reception and to keep the TV color oscillator locked on a broadcast station.

C Symbol for chrominance

Capstan A small magnetic tape drive shaft which drives the recording tape to assure positive tape movement.

Capstan Servo An electro-mechanical system that maintains capstan motor rotational speed in sync with a comparison signal generated within the electronics of the servo system. A self-correcting speed control system.

Carrier A waveform containing signal information.

Carrier Leak A beat effect seen on a display screen as a series of wavy lines. This is a phenomenon that occurs whenever a plain carrier signal appears during playback of a video signal recorded as FM.

Channel A specified range of frequencies used to transmit and receive TV signals.

Chroma The color part of a video signal. Describes the degree of saturation of color. Dark red is saturated; it becomes less saturated as white light is added to produce light red. On a standard TV, chroma is adjusted with the "COLOR" control knob.

Chroma Bandwidth The frequency difference between the upper and lower frequency limits of the chroma signal.

Chroma Delay A time delay produced by restricting the chroma band of frequencies through a low pass filter.

Chrominance Another term for chroma.

Chrominance Signal The composite signal produced by combining hue and chroma. The chrominance subcarrier signal conveys hue and saturation information. (See luminance.)

Clamp Holding an AC signal at a specified DC level.

Clogging The blocking of the head gaps by dirt or oxide abrasion from the surface of the tape. Clogging causes noise and an inability of the VCR to perform properly.

Coercive Force The amount of opposing magnetic intensity that must be applied to a video tape to remove (erase) residual magnetism.

Coercivity A tape property related to the amount of coercive force required to demagnetize a magnetically saturated portion of tape material.

Color Bar A color test signal that contains bars of color and peak black and white. This set of bars is used to determine luminance and chrominance. From left to right, the bars are white, yellow, cyan, green, magenta, red, and blue. Seventy-five percent of the peak level of this standard color test signal is the input level for any of the primary colors.

Color Difference Signals The resulting signals after subtracting the composite luminance Y signal from the primary color signals red, green, blue, red-Y, blue-Y, and green-Y.

Color Flicker A 30 Hz change in color saturation caused by video preamp imbalance between two video heads.

Color Signal A composite signal comprised of luminance, the chrominance subcarrier, and a color sync signal.

Color Sync Signal Also called Burst and Color Burst. Consists of 8 to 12 cycles of chrominance 3.579545 MHz subcarrier that are added to the end of the TV horizontal sync signal. This signal becomes the reference for phase comparison during reception.

Color Temperature A way to express color in terms of absolute temperature. A black body heated to high temperature emits light. The relationship between color of light and temperature is constant. Bluish colors have a high color temperature; reddish colors have a low color temperature.

Color-Under A color-recording technique in which

chrominance information is converted (heterodyned) from 3.58 MHz to 688 kHz.

Comb Filter An electronic circuit with a frequency pass response resembling a comb. It consists of a 1 H delay line and a summation point that adds (or subtracts) interlaced video to separate out the luminance (or chrominance) signal.

Compatibility The ability of a tape recording made on one VCR to be played back on another and vice versa.

Composite Sync Signal The combination of video, blanking, and composite sync signals.

Condensation Moisture deposits on the head drum will cause the recorder to stop operating until the drum is dry. Operating a VCR when condensation is on the drum can cause the tape to stick to the drum and be ruined. Some recorders have an illuminated moisture indicator.

Control Signal A special signal recorded on a video tape that is used during playback as a reference for the servo circuits. This pulse is recorded on the Control Track and marks the beginning of a recorded video track and is used during playback.

Control Track A special video tape track used to store tape speed synchronization timing control signals.

Converted Subcarrier The process of shifting the 3.58 MHz color subcarrier and its sidebands down to 629 kHz for VHS (688 kHz for Beta).

Counter EMF An electric field force that is opposite to the field that originated it.

Crosstalk Unwanted signals detected by a video head from an adjacent track.

CW Continuous Wave.

CW Interference RF interference in a waveform that is generated by an external carrier signal.

Dark Clip A circuit used to cut off large negative voltage spikes at an adjustable level to prevent damage to the FM modulation circuitry.

dB Decibel. A unit used to express ratio of power or sound pressures.

DC Error Voltage A comparator output proportional in magnitude and polarity with the operational error in a feedback control system.

DDC Direct Drive Cylinder. A video head cylinder driven by a self-contained brushless DC motor using no belts or gears. Use of DDC produces increased stability pictures.

Decoder A device that converts a coded signal back to its original uncoded condition.

De-emphasis The process of attenuating previously amplified signals to produce the original signal with less noise.

Definition A unit of measurement for the sharpness of an image. Denotes the degree of distinguishable picture details. Definition is divided into vertical and horizontal.

Delay Multivibrator A monostable multivibrator that generates an output pulse a predetermined time after being triggered by an input pulse.

Delta Factor Used to indicate a change in frequency, such as jitter or "wow and flutter" in a playback signal where the color signal off the tape is not frequency stable at 629 kHz, but varies about the 629 kHz point.

Demodulation The conversion of high frequency modulated TV signals to audio or video.

Deviation Used to describe the FM carrier frequency swings during modulation. The excursion of an FM carrier from the original represents the information signal.

Direct Recording A videotape recording technique in which the complete composite signal is processed and recorded.

Dropout A momentary absence of FM or color signal on a tape where no video signal was recorded. Dropout is caused by uneven tape oxide, dust on the tape or video heads, or mechanical damage. It distorts the TV picture and appears as tiny white dots or as white streaks.

Drop Out Compensator A device that electronically fills in RF waveform holes produced by dust and tape imperfections which prevent accurate playback. The circuit uses the RF information from the previously stored line.

Drum Servo The system that controls head rotation by comparing the phase of the vertical sync signal with a signal produced by drum rotation. Any difference is detected as error voltage and is amplified and applied to a brake coil to modify rotational speed.

Duty Cycle The percent of a waveform that is active during a single cycle of the period.

Eddy Current A circular flow of electrons in a conductor exposed to a rapidly changing magnetic field. Eddy currents increase with frequency.

Editing Adding, deleting, or rearranging sections of audio or video recorded on a tape.

E-E Electric-to-Electric, Electronic-to-Electronic. A condition during record where the playback output circuit is connected directly to the record input circuit so the video signal can be monitored on a TV set for proper tuning.

Effective Video Width The amount of tape width occupied by video recording less the overlap.

EIA System The TV broadcast standard established by the Electrical Industries Association that uses 525 horizontal scanning lines and 30 frames per second to produce a TV screen image.

Eject Mechanically or electrically opening the cassette housing of a VCR so a video tape cassette can be removed.

EMF Electro-motive Force. The electric force produced by moving electric or magnetic fields.

Emphasis A process for boosting the level of high frequency components in a video signal.

End Sensor An automatic stop device using metal foil tape leader and two contactless oscillator coils for Beta systems or an LED (or incandescent lamp) and sensing device with clear leader tape for VHS systems.

Entrance Guide A tape guide mounted at the input side of the scanner to control tape angle and height.

EP Extended play (same as SLP).

Equalizing Pulses Six serial pulses at twice the line frequency that occur before and after the vertical sync signal to ensure proper interlaced scanning and to retain horizontal sync during blanking.

Error-In A servo feedback signal that corresponds to the deviation from a reference standard.

Factor 80 A heuristic relationship between lines of resolution and bandwidth: Bandwidth X 80 = No. lines of resolution.

FG Frequency generator, circuit that generates a square or sine wave signal corresponding to the rotational speed of the video head drum, capstan, and reel tables.

Field The amount of picture that can be described during a vertical scanning period of one sixtieth of a second. It takes 15,750 monochrome horizontal scanning lines (15,734 for color) and 60 fields to draw an image on a screen in a single second (U.S., Canada, and Japan). Two vertical scans will scan 525 lines of horizontal scan so 30 pictures can be drawn each second. (525 lines = 2 vertical scans = 1 frame X 30 frames per second = 15,570 monochrome horizontal lines/second).

Field (Flux) An electric or magnetic force that exists between two unlike poles or around a single pole that can be represented by lines of force flowing out of one pole and into the other.

Field Intensity A measure of the magneto motive force per unit length. Expressed in oersteds and symbolized with the letter H.

Field, Magnetic The force field around a magnetic particle.

Field Rate The 60 Hz rate at which a field (half screen image) of 262.5 lines appears on the screen.

Field Skip A system in which only one or two video heads records every other field on tape. During playback, the first head traces a track, and then the second head traces the same track.

Flip-Flop A collection of logic gates configured such that a logic High input will produce a change of output from High to Low or vice versa.

Flux The summation of magnetic field effects through an arbitrary space. Expressed in maxwells.

Flux Density A measure of the flux per unit area. Measured in gauss and symbolized by (β).

FM Demodulator Carrier frequency doublers and pulse density detector circuits which recreate original video from a modulated carrier through use of a low pass filter.

FM Recording The modulation of a carrier wave in accordance with the amplitude of a video signal. FM recording permits recording of low frequencies and eliminating level fluctuations.

Frame Rate The 30 Hz rate at which a frame (525 lines) appears on a video screen.

Frequency A measurement of the number of electromagnetic cycles or alternations that occur in one second.

Frequency Interlace A method to interweave chrominance (color) and luminance (brightness, black and white) information within the same bandwidth.

Frequency Modulation The variance of a carrier wave frequency in direct relationship to an input signal.

FWD Search Forward search mode.

Full Field Recording Unlike skip field recording, this technique continuously uses both of the two video heads for both recording and playback of every field.

Full Logic Transport Control Circuitry to enable smooth transition between VCR operating modes. Enables the use of remote control and automatic electronic editing.

Gap The space between the two poles in a magnetic tape head.

Gauss The CGS unit for measuring flux density.

Gilbert The basic measuring unit for magneto-magnetic force.

Ghost A vague image reflection appearing to the side of a main TV image caused by antenna reflection or impedance mismatch on the antenna producing a reflected wave of signal.

Guard Band An empty space of tape between two video tracks that provides enough distance between the two tracks to prevent crosstalk. The difference between the video track pitch and the video track width.

HAFC Horizontal Automatic Frequency Control (system). To automatically maintain the horizontal oscillator frequency within set limits.

HD Horizontal Drive. A sync pulse used to control the horizontal drive.

Head Cylinder The metal cylinder which houses the video heads.

Head Drum A mounting plate containing video heads which rotates to achieve high writing speeds.

Head Gap The space between the two pole pieces of a ferrite core used as a video head. Typically, this space is filled with glass to maintain the critical spacing dimension.

Head Switching The process of switching between video heads in a multiple head system.

Head Switching Pulse A 30 Hz square wave that is applied to the Head Amplifier to perform head switching.

Helical Scanning A technique for video recording in which the heads rotate horizontally on a moving drum while the tape is drawn around the drum and over the heads diagonally. This produces diagonal recording tracks on the tape. This recording technique makes mechanical and electronic construction simpler and the system easier to maintain.

Helix Angle The angle between the plane in which the video heads rotate and the edge of the tape.

Horizontal Blanking The period in which the electron gun is off during horizontal retrace. Horizontal sync is 4.76 usec; blanking takes 10.4 usec.

Horizontal Resolution The number of horizontal pixels that can be distinguished on a horizontal scanning line. Commonly described as "lines of resolution."

H Sync Horizontal sync.

Hue The characteristic of a color that makes it unique (e.g., red, yellow, green, blue, etc.). Determined by the dominant wavelength in a color signal regardless of saturation. Represented by the phase angle in NTSC systems.

Hysteresis The relationship between flux density and magnetic field intensity wherein a lag occurs in the magnetization value due to the changing magnetic force.

Hysteresis Motor A synchronous motor that is driven by a permanent magnet rotor in a rotating field.

Induction The phenomenon in which two materials interact electromagnetically due to proximity. The existence of a magnetic field in one material establishes a magnetic field in another adjacent material.

Induction Motor An A.C. motor in which current flow in the primary coil induces current flow in a secondary coil wrapped around the armature.

Interchangeability A description of how well a tape recorded on one VCR will play on another VCR.

Interlace A scanning system in which the lines of each frame are divided into two fields and half are scanned (odd) then the other half (even) in alternate fashion to generate a 525 line picture or frame in one thirtieth of a second.

Interlaced Scanning The scanning of 262.5 odd lines followed by the scanning of 262.5 even lines to produce a complete picture frame. This method reduces flicker without the need to broaden the video signal bandwidth.

Interleaving A condition in which the harmonics of the chrominance signal occurs in between the harmonics of the luminance portion of the video signal. Interleaving of color information prevents interference with luminance information although it is broadcast at the same time.

Intrinsic Coercivity The property that determines the level of field intensity at which saturation occurs.

Jitter The instability in a playback signal caused by tape or head speed fluctuations. This excessive wow and flutter causes the picture to have a rapid shaking or shivering movement.

Limiter A circuit that produces a constant amplitude output regardless of varying input. A clipping circuit to strip off amplitude variations in an RF waveform to produce a frequency modulated signal.

Loading The mechanical operation from cassette insertion to completion of tape path set-up. In VCR cassette systems, an automatic loading system takes the tape out of the cassette and wraps it around the head drum and capstan so recording and playback can occur.

Lock-Phase Delay A timing circuit that delays the PGA pulse to make it coincident with the on-tape vertical sync signal.

LP Long play.

Luminance A signal containing brightness information. In black and white broadcasts, only the luminance signal is transmitted. Symbol Y.

Magneto-Motive Force mmf. The force of a magnetic field of one mass particle on another mass particle.

Magnetization Inducing a residual flux in a material which is proportional to an applied magnetic field intensity.

Magnetic Domains The intrinsic charge distribution of matter. Charges are divided into balanced pairs or dipoles.

Magnetic Moment The force exerted at the axis of each magnetic domain or dipole by an external field.

Markers Dropouts that are placed in a video sweep as reference points for alignment.

Maxwell The basic unit of measurement for a line of force.

MMV Monostable Multivibrator. A circuit which produces a logic high or low output with variable duration in response to an input pulse or voltage level transition.

Modulator A component or circuit which converts audio and video signals into a television signal by varying the amplitude, frequency or phase of a carrier as a function of an input signal. Some quality loss in the conversion is unavoidable.

Monitor A television set without a tuner section.

Monoscope A video display test pattern used for rough evaluation of resolution and SNR.

Monostable Multivibrator See MMV

Noise Bar A rolling horizontal bar on a display screen caused by video heads passing over nonrecorded guard bands.

Noise Canceller Circuit A circuit designed to remove cross color noise from harmonics of the sync and luminance signals.

Nonsegmented A video recording system in which a complete field is recorded for each pass of a video head.

Nonsynchronous Motor A motor with a rotational speed less than the line frequency. The conductor bars of the secondary winding cut the rotating flux field.

NTSC National Television System Committee. The present color telecasting system in the U.S., Canada, and Japan which uses 525 lines and 60 frames per second to produce a visual image on a TV screen.

Oersted The unit measure of magnetizing force.

Overlap The condition in which the beginning of head B output overlaps slightly with the end of the head A output eliminating any blank areas between the two tracks. Recorded video is reproduced by the alternating output of the two video heads.

Omega Wrap A tape threading method in which the tape wrap around the head drum resembles the Greek letter omega.

OR Gate A logic gate in which any input of a logic High will produce a logic High output.

Overmodulation The condition in which a large signal input to the FM modulation circuit increases the oscillator frequency to such an extent that a black edge-like effect is seen on the right side of white portions of the screen image.

PAL Phase Alternation Line. The color television system used in Western Europe (except France) which uses 625 lines and 50 frames per second to produce a visual image on a TV screen.

PAL-SECAM Adaptor A special attachment to a PAL TV set which permits the receipt and display of SECAM color TV programs.

Pause Temporarily stopping the tape transport while remaining in the same mode (PLAY or RECORD).

PCM Pulse code modulation. A signal modification technique used in digital audio recording and playback (Sony Beta and 8 mm).

Permeability The ability of the magnetic domains in a material to align under the influence of an external magnetic field. Permeability is symbolized by μ.

PG Pulse Generator. The circuitry that produces the rectangular wave pulse signal that is used as a reference during playback and as a comparison signal during recording. A PG pulse is generated each time a video head rotates past a PG coil. With two video heads, a positive pulse is produced on the approach and a negative pulse is produced as the magnet leaves the coil. The two signals are then combined into the rectangular square wave.

PGA, 30 A timing pulse generated by head drum rotation and used as feedback in the scanner servo.

PGA Phase Delay introduced by the 30 PGA and PGB multivibrators.

PGB, 30 A second servo timing pulse generated near the head drum for timing video head switching.

Phase Comparator A circuit that compares phase timing of two events to produce a feedback correction voltage.

Picture Instability A temporarily unstable display caused by irregular tape travel and its resultant sync signal interference.

Picture Scanning The method in which a television picture is drawn on a screen using two fields scanned line-by-line, top left to bottom right. The number of lines determines the quality and detail present in the produced image.

Picture Search Cue and Review, Shuttle Search, Visual Search. A condition in which the recorder operates in fast play without sound so the operator can locate a particular point on the tape by sight rather than by counter.

Pitch The distance between adjacent video tracks which is determined by tape speed.

Pole, Magnetic One of two regions (dipoles) of a magnetized body at which flux density is concentrated.

Pre-equalization The process of boosting the low frequency components in a video signal before processing to compensate for low frequency roll off during playback.

Pre-emphasis The boosting of frequencies to compensate for future frequency roll off and to reduce noise by increasing the SNR.

Pulse Generator A device in the video head servo sync system to produce timing pulses as a head drum rotates.

Rabbet The shelf of the lower video head drum surface which guides the tape around the scanner.

Raster The 525 horizontal lines which comprise a complete video frame (picture).

Reference Video Head One of two heads in a helical scanner which is chosen as a frame of reference.

Relative Speed The speed relationship between the tape and the rotating head. High frequency video cannot be recorded at audio tape speeds so the head was made to rotate relative to the tape which greatly increased the effective speed of the system.

Reluctance The resistance of a material to magnetic influence. This is the inverse of permeability.

Remnant Magnetism The flux remaining in a material after an external magnetic field has been removed.

Residual flux The flux associated with remnant magnetism.

Resolution The ability to reproduce image detail. When a black and white pattern of lines is produced on a screen, the number of visible lines is used to express the degree of resolution.

Retention The ability of a magnetic material to hold a residual flux pattern after an external magnetic field has been removed.

REV Search Reverse search.

RF Radio frequency. Any frequency at which coherent electromagnetic radiation of energy occurs.

RF Envelope The curve formed by the peaks of a signal at the video heads or "off-tape" video.

RF Sweep A 1 field sweep signal containing frequencies from zero to 7 MHz and dropout markers at 1, 2, 3.58, 4.5, 5.4, and 7 MHz.

RF Modulator A device that converts audio and video signals present in a VCR to an appropriate TV broadcast signal. Usually the VCR produces TV signals at television broadcast Channels 3 or 4.

RF Transmission The method used to transmit video signals by converting the audio and video information to television broadcast signals and passing this information through a cable to the television receiver. The signal is converted to the frequencies associated with an unused TV channel.

Rotary Chroma The VHS process that changes the phase of the chrominance signal at 15.734 kHz.

Rotary Transformer A device that magnetically couples RF signals to and from a spinning video head. This device eliminates the need for brushes in a VCR.

Rotating Drum The cylinder around which the tape is pulled. In helical scan, the tape traces a diagonal path around the drum forming a helix so that the rotating heads can trace a diagonal path across the tape surface.

Rotating Head A head (or heads) mounted on a drum cylinder that rotates horizontally as a tape is pulled past on a diagonal path. Over 200 times audio tape speed is achieved by rotating the heads at a high speed while maintaining a slow tape movement. This technique realizes high relative speeds with greatly increased recording area and more efficient use of tape surfaces.

Sample-and-Hold A comparator circuit process in which a particular signal is measured (sampled) at a specific moment in time and held for later use.

Saturation The point at which all the magnetic domains in a material are aligned and a further increase in magnetic intensity does not result in greater flux density. Color saturation is the depth of a color or spectral purity as determined by the relative amplitude of the chrominance signals.

Scanning The process of decomposing a video image into discrete lines of brightness and color (hue and chroma) information and then reconstructing the same image on a television screen. Both horizontal (left to right) and vertical (top to bottom) scanning occur simultaneously.

Scanner Servo A feedback control system used to adjust video head timing by controlling head drum speed.

SECAM Sequential Couleur a Memoire. The 625 line, 50 frames/second color television standard used in France and East Europe.

Segmented A VCR system in which each video head pass across the tape deposits a fraction of a field and more than a single pass is required to record a complete field.

Servo An electro-mechanical device whose mechanical section is constantly measured and regulated by the electrical section so the drive speed or position of the mechanical section closely follows an external reference.

Set-up The difference between the black level and the

blanking (no video) level and expressed as a ratio of the black/blanking magnitude to the blanking/peak white magnitude.

Sideband Cutting The process of limiting bandwidth of one side of a sideband pair by attenuating one sideband to unbalance the pair.

Skew The change in size or shape of video tracks from the time of recording to the time of playback. When present, this usually results from poor tension caused by misalignment or ambient conditions that affect the tape or tape path. The problem appears as a TV image that is bent out of shape.

Skew Error An error describing horizontal sync and video signal advance or retard from normal timing.

Slip Ring Metallic rings that are connected to the coils on the rotating heads and which form the interface with the heads and the electronics of the VCR.

Slow Motion Generating screen images at slower than normal speed.

SLP Super Long Play.

S/N Signal-to-Noise. A measure of the ratio of the desired signal to unwanted noise, expressed in decibels (dB). The higher the ratio, the clearer the screen image and the higher quality the audio. Video S/N is about 40 dB.

SP Standard play.

Standard Tape A high tolerance reference tape used to evaluate machine performance during design and prototyping of a new VCR.

Subcarrier The 3.58 MHz continuous wave (CW) signal used to carry color information.

Suppressed Carrier A color recording technique in which the carrier is attenuated or cancelled and only color sidebands are produced.

Sync Head A head in a video recording system that records information only during the vertical blanking periods. Called a "1½ head."

Synchronous Motor A motor whose speed is synchronized to the line frequency.

Sync Signal The pulse signals that synchronize the video circuitry. Horizontal sync defines the start of horizontal scanning; vertical sync determines the beginning of vertical scanning.

Sync Tip The lowest point in a video waveform corresponding to the horizontal sync signal.

Tape Guide A metal cylinder or tapered metal shape that guides the tape properly past the heads, capstan and along the tape path.

Tension Servo A servo system that maintains constant tension along a video tape between the supply and take-up reel spindles.

Tesla The unit for measuring flux density.

Threading The operation in which a cassette tape is automatically positioned around the tape path.

Time Base Errors Mechanically induced timing errors in a VCR caused by incorrect tape tension, tape stretch, etc.

Time Base Stability Term used to describe how closely sync timing enables the playback video signal to match an external reference video signal. The ability of a VCR to minimize recorded errors due to mechanical transport errors that cause hue errors from frequency and phase deviation of the color signals.

Timer A device built into the tuner of a video section that enables the recording of different TV broadcasts at preselected times and days.

Timing Error (Phase) The deviation from the correct rotational position of the head drum in reference to a CTL pulse.

Track The magnetic pattern imposed into a tape surface by a recording head.

Track Angle The angle between the video track and the edge of the tape.

Track Length The tape distance on which video information is recorded.

Track Width The core piece thickness at the head gap which controls the width of the tape area affected by the magnetic field from the head gap area.

Tracking The ability of the spinning video heads to accurately pass over the recorded video RF information during playback. Good tracking is indicated by a strong RF signal; poor tracking is indicated by low level RF or noise.

Tracking Adjustment An electronic adjustment that delays the CLT (Control) signal in a playback VCR so its effective position matches the position of the CTL head on the recording machine.

Tracking Control A variable delay MMV used for tracking adjustment.

Tracking Errors The visible interference noise bars produced on a screen caused by the inability of video heads to correctly follow the video tracks recorded on a tape. The tracking control adjustment corrects for this error.

TTL Transistor-transistor-logic.

U-Loading A tape system in which the tape forms a U shape as it passes around the head drum.

U-Matic A video system standard.

Unloading Replacing the tape back into its original position in the cassette.

Utilization Ratio The ratio of lines in a raster contributing to vertical resolution and the total number of raster lines. This ratio should be 0.7.

VCO Voltage Controlled Oscillator. An oscillator whose frequency is controlled by an an external voltage level.

Vertical Blanking The video electron gun off-period during vertical retrace. Vertical sync is 3 horizontal lines long while blanking is 21 lines long.

VHS Video Home System. A tape recording format developed by JVC in Japan and in use worldwide.

Video Drum A drum cylinder with opposing video heads mounted 180 degrees apart.

Video Head An electro-magnet wrapped at one end with a tiny coil of wire in which a signal current flows producing a magnetic flux field that enables RF information to be stored on a tape.

Video Muting The electronic blanking of a VCR screen to black during machine threading and unthreading, start-up and playback.

Video Track The area on a tape in which RF information is magnetically recorded as a recording head passes over.

Visual Color Acuity The ability of the human eye to discern different shaped colors at varying distances. The ability to resolve small picture shape and color details.

V Sync Vertical sync.

VTR Video Tape Recorder. A collective term for all video recording systems.

V-V Video-to-Video. The playback screen image produced from a tape.

VXO Voltage Controlled Crystal Oscillator. An oscillator whose frequency is controlled by an internal quartz crystal.

Weber A measure of flux. 1 Weber = 1 X 10*8 Maxwells.

White Clip A circuit used to cut off excessive positive-going spikes (overshoot) from the emphasis circuitry and hold the overshoot to a predetermined, adjustable level.

Wow and Flutter Tape transport speed fluctuations that can cause regular instability in the screen image and produce a wavering or quivering effect in the sound during recording and playback. Fluctuations below 3 Hz are called "wow," and short cycles above 3 Hz are called "flutter."

Wrap Angle The angle produced by the tape against the scanner surface and the center of the scanner.

Writing Speed The relative speed between the record head and the tape surface.

XTAL Crystal

Y Signal The portion of a video signal containing black and white information and sync.

Bibliography

Advokat, Stephen. "Have One VCR? How About Two?" *Times-Advocate,* May 11, 1986, 4.

Allen, David. "Beta Gets Better." *Videography,* June 1985, 79-80.

Babcoke, Carl. "Tips for Better Cassette Recordings." *Electronic Servicing & Technology,* June 1983, 22-39.

Basch, Vladimir. "Finding a Surge Suppressor You Can Trust." *Computer/Electronic Service News,* July 1985, 60-69.

"Battle Lines Are Drawn in War of Video Formats." *The Institute,* August 1986, 6.

Bentz, Carl. "Reception Problems? Take a Look at the Coaxial Cable." *Electronic Servicing & Technology,* June 1983, 53-55.

Bentz, Carl. "TV Signals Around the World." *Electronic Servicing & Technology,* November 1985, 44-51.

Betts, Kellyn S. "Magnetic Media The Memory Lingers On." *Modern Office Technology,* July 1985, 78-86.

Bowden, Steve. "VCR Basics." *Electronic Servicing & Technology,* February 1984, 46-51.

Bowden, Steve. "VCR Basics - VHS Servo Operation." *Electronic Servicing & Technology,* October 1984, 42-50.

Bowden, Steve. "VHS Basic Recording and Playing." *Electronic Servicing & Technology,* June 1984, 41-48.

Bowden, Steve. "Videorecorder Basics." *Electronic Servicing & Technology,* March 1984, 20-24.

Brenner, Robert C. *IBM PC Troubleshooting & Repair Guide.* Indianapolis: Howard W. Sams & Co., 1985.

Butler, Robert. "Uninterruptible Power Supplies." *Electronic Servicing & Technology,* June 6, 1985, 181-186.

Clifford, Martin. "The Video Connection." *Electronic Servicing & Technology,* March 1985, 52-55.

Clifford, Martin. "The Video Connection." *Electronic Servicing & Technology,* October 1985, 50-55.

"Close-Captioned Recording." *Changing Times,* August 1986, 78.

Consumer Reports 1986 Buying Guide Issue. Mount Vernon, NY: Consumers Union of United States Inc., 1986.

Dobbin, Peter. "The CES Video Picture." *High Fidelity,* April 1985, 14-26.

Doherty, Rick. "JVC Rolls Out Better Hi-Res VCR Format to Rival Beta Systems." *Electronic Engineering Times,* October 7, 1985, 53.

Doherty, Rick. "Sony Previews New TVs, VCRs." *Electronic Engineering Times,* May 5, 1986, 17.

"For Home Video, Prices Have Hit Bottom, and It's Bye-bye for Beta." *The Institute,* March 1986, 8.

Foster, Edward J. "Is Automatic Better?" *High Fidelity,* February 1985, 31-82.

Goodman, Robert. "An Ounce of Prevention." *Electronic Servicing & Technology,* May 1983, 24-39.

Graham, Wayne B. "Diagnosing VCR Head Problems." *Electronic Servicing & Technology,* September 1985, 22-61.

Haldin, Ken. "The Coming of Age of the Videotape." *Los Angeles Times Home Entertainment Guide,* November 20, 1985, 4.

Heller, Neil R. "Servicing Videocassette Recorders, Part 1." *Electronic Servicing & Technology,* August 1985, 46-53.

Heller, Neil R. "Servicing Videocassette Recorders, Part 2." *Electronic Servicing & Technology,* September 1985, 12-20.

Heller, Neil R. "Servicing Videocassette Recorders, Part 3." *Electronic Servicing & Technology,* October 1985, 37-43.

"High-Tech Teens." *Wall Street Journal,* July 17, 1986, 25.

Kybett, Harry. *Video Tape Recorders.* Indianapolis: Howard W. Sams & Co., 1986.

Benk, John D. *Complete Guide to Modern VCR Troubleshooting and Repair.* Englewood Cliffs: Prentice-Hall, Inc., 1985.

Long, Robert. "Audio/Video Crosstalk." *High Fidelity,* January 1985, 21.

Long, Robert. "What to Do 'til the Repairman Comes." *High Fidelity,* July 1985, 22-23.

Matsushita Training Manual Technical Descriptions Models PV1730, PV5800/VP5442XQ, PV8500k/VP5740XQ, PK950/VK744XE. Matsushita Electric Corporation of America, 1984.

Mitchell, Steve. "Expert Advice on Purchasing a Videocassette Recorder." *Los Angeles Times Home Entertainment Guide,* November 20, 1986, 5.

Mowrer, William. "Video for a New Age." *High Fidelity,* September 1984, 50-51.

Panasonic Service Manual NV1300 Vol. 1. Matsushita Electric Corporation of America, 1979.

Panasonic Service Manual PV1000A Vol. 1. Matsushita Electric Corporation of America, 1976.

Persson, Conrad. "What's New in Video?" *Electronic Servicing & Technology,* June 1983, 56-58.

Pike, Helen. "New Technologies Could Change the Face of Television." *Electronic Engineering Times,* November 4, 1985, 32.

RCA The L Line Video Cassette Recorder Service Education. RCA Consumer Electronics, 1985.

"Replacing Upper Head Cylinder in RCA VCRs." *Electronic Servicing & Technology,* December 1984, 42-47.

"Restoring Versatility to Your Cable-Connected VCR." *Electronic Servicing & Technology,* February 1985.

Rice, John. "Audio for Video." *Videography,* June 1985, 84,85.

Riggs, Michael. "Basically Speaking - Horizontal Resolution." September 1984, 22.

Riggs, Michael. "What You Can (and Can't) Hear." *High Fidelity,* November 1985, 28.

Shmatovich, Chris. "Connector, Cable Crosstalk in Controlled Impedance Cable Assemblies." *Electronics,* August 1984, 51-54.

SL8200 Circuit Manual and Mechanical Description. Sony Corporation of America, 1977.

SL-HF300 Operation Manual. Sony Corporation of America, October 1984.

SL-HF900 Operation Manual. Sony Corporation of America, November 1985.

"Special: CBS Labs Test Blank VHS Tapes." *Video Review,* September 1985, 50-71.

Sony Service Training SL8600. Sony Corporation of America, 1978.

Sony Training Manual SL5200 and SL2400 Course B-04. Sony Consumer Service Company, 1983.

"STANDARDS Digital VTR Standard Imminent." *IEEE Spectrum,* October 1985, 30.

"Survey: VCR Boom Likely to Continue." *San Diego Transcript,* April 11, 1986, 7.

Sweeney, Dan. "Videocassette Recorders Are Not All Created Equal." *Video Times,* June 1985, 14-16.

The Sams Hookup Book. Indianapolis: Howard W. Sams & Co., 1986.

"Top-Flight VCRs." *High Fidelity,* May 1985, 36-39.

Training Manual Portable Video Cassette Recorder, Technical Descriptions PV5200/5500. Matsushita Electric Corporation of America, 1982.

"VCR Troublshooting." *Electronic Servicing & Technology,* May 1985, 21-28.

Video Service Data: Videocassette Recorder Quasar VH5040SW. Indianapolis: Howard W. Sams & Co., 1982.

Vistain, Kirk. "Audiocassette Recorder Adjustment." *Electronic Servicing & Technology,* December 1983, 20-53.

"Vital Statistics." *TV Guide,* June 21-27, 1986, A-133.

"What's the trouble?" *Electronic Servicing & Technology,* July 1983, 4.

Whitaker, Jerry. "Don't Let Power Line Disturbances Damage Your Electronic Equipment." *Electronic Servicing & Technology,* February 1985, 38-63.

Wolfe, William. "Spec Speak." *Video,* October 1985, 102-151.

Zenith Service Manual Vol. 3 Theory of Operation. Zenith Radio Corporation Parts and Service Division, November 1980.

Appendix

SELECTING A VCR

Selecting a video tape recorder can be a frustrating and confusing experience. Here are some guidelines to help you make your selection.

First, understand clearly how the VCR will be used. Will it be used during tape exchanges with a neighbor or relative? Will you use it just to play back rented movies on the weekend, or to record programs while you are away from home? Consider what your immediate and future needs might be so the VCR you select will be appropriate now and later. Ask friends or relatives who own a VCR how they feel about their particular machine and its options. Ask what they wish their VCR could do, what options they will buy on the next machine.

Selecting a VCR format is important too. Technically, the Beta format design produces superior video, but VHS is far more popular. However, the recent introduction of SuperBeta has given the world a new standard of excellence. Once again, Sony has leap-frogged to the technical forefront of video recording and playback capability. The guard band design of Beta and now Super Beta results in video reproduction with minimal dropout and sharp, crisp images. Unquestionably, Beta excels in picture quality. VHS, however, has been widely received in the United States. VHS machines excel in recording time capability.

You must decide if you want Beta, VHS, or 8mm. You may wish to base part of the decision on the kind of tapes available for rent at your local video store. Consider what format your friends and relatives are using in case you want to exchange tapes with them.

Do you intend to use the VCR to record home movies? Do you receive TV programs through a cable system or an antenna? If antenna, do you intend to get cable in the future? Do you want special effects features such as slow motion, or a fast playback speed which allows you to play the program at two or three times the normal viewing speed in either forward or reverse. Or perhaps you'd like a still frame feature that allows you to pause the machine during playing so that you have a perfect still picture (similar to a slide) on your TV screen. Then there are the search, or cue and review features that allow you to zip through unwanted sections of your program either in forward or reverse at about nine or 10 times the normal playback speed. Cue and review is one of the most popular features for reviewing a particular scene or speeding through commercials in one-tenth the time.

Many machines allow you to pause during playback with some form of a viewable picture on the screen, and to search in forward or reverse without offering other, more expensive special effects. Good still frame and slow motion are features generally reserved for the more expensive machines.

The type and number of programming events that can be set will determine if you will be able to record one or more programs while you are away. The key feature is the timer. A VCR timer lets you record a program automatically. The recorder is preprogrammed to turn on to a specified channel at a selected time, to record for a set period of time, and to turn off automatically at the end of this period.

The basic timer can record one program within a 24-hour period. More expensive models let you preprogram two, three, four, or more programs over a period ranging from a few days to several weeks. On some VCRs, you preprogram events up to a year in advance. Selecting special effects generally creates the most confusion during a VCR purchase.

Another available option is the tuner itself. Some VCRs have what is known as a cable-ready tuner. Since the video industry has no formal standard defining cable-ready tuners, you might find the following general discussion to be helpful.

Standard TV broadcast signals consist of three bands of frequencies—VHF low band, VHF high band, and UHF. Cable companies transmit their signals on the VHF low band, VHF high band, and on special bands known as mid-band, or super-band.

Other bands such as the hyper-band are also used occasionally but the mid- and super-bands are most commonly used. Cable companies don't use the UHF band. Many VCRs require the use of a special cable box to receive mid- and super-band channels. These channels are converted to a VHF low frequency band such as channel 3 so that a normal TV can receive the programs. The cable boxes receive and tune the mid-band stations.

Some VCRs come with a cable-ready tuner, which means the tuner is designed to directly receive the cable transmitted mid- and super-bands.

This eliminates the need for a cable converter box. The cable converter box is required if you want to receive scrambled pay-TV cable programs such as "HBO," "Disney Channel," or others.

Another factor which may arise in the selection of your VCR is the advantage of direct drive motors instead of a belt driven mechanism. A direct drive motor VCR means that each essential mechanical, moving part of the VCR is directly driven by its own separate motor without using belts, pulleys, or gears.

Most VCRs have more than one motor. For example, one motor may be required to load and unload the tape while another is used to move some other part of the machine. In a front load machine, a separate motor is used to raise and lower the video cassette into the unit. Separate motors are also used to rotate the head drum, and the capstan and reel tables.

In most cases VCRs with direct drive motors and no belts are more expensive. But VCRs with direct drive motors provide extra advantages such as longer performance life (no belts to replace), and a much tighter control over tape handling. This provides finer edits, better tape control and, in general, better stability of video recording and playback.

On the other hand, direct drive motors require more electronic circuitry for precise operational control. Another disadvantage is that if these motors fail, the repair can be more expensive. There are typically more motors in VCRs with direct drive (cassette tray assembly motor, threading motor, capstan motor, head drum motor, take up reel table motor, supply reel table motor, etc.).

Original VCRs were driven via a series of belts, pulleys and idler tires. These machines had only two motors. One motor was responsible for the action of the threading, and another motor controlled all other VCR operations. The latter (main) motor was powered directly from the 110 volt AC line. It rotated at 1800 revolutions per minute (RPM) and was synchronized by the 60 Hz AC input. The head drum was rotated at a precisely controlled speed as calculated by the diameter of pulleys over which the connecting belts were attached. The capstan shaft was rotated at a precisely controlled speed by another series of belts and pulleys connected to the same AC motor. Reel tables were also driven by this same AC motor through a series of belts, torque limiter assemblies, tires and pulleys.

With the introduction of two- and three-speed machines, a different motor system was required to allow for changes in speed. This lead to the introduction of the DC controlled capstan motor. The head drum and reel tables were still controlled by the AC motor. In current technology VCRs we find three-motor belt-driven machines. These motors include the loading motor, capstan motor and drum motor.

Belt-driven cassette recorders are commonly found among the middle and lower priced VCRs. Manufacturers recommend that belts and rubber tires be replaced every one to two thousand hours of use. Belts can break without warning, and tires may begin to slip even if the unit was not used excessively. Rubber will deteriorate with time causing possible defects. Defective belts and tires can cause the unit to not play, to slow, or to not rewind or fast forward. Such defects can cause excess tape to remain in the machine after the cassette is ejected.

A lot of thought should also be given to the sound features you want. VCRs now have hi-fi sound or high definition audio. Hi-fi and high definition audio require special electronics and, in the case of VHS, extra heads must be mounted on the video head drum assembly. These machines reproduce stereo sound with fantastic quality, exceeded only by compact disc players. External amplification and external speakers are required to use these features. These sophisticated VCRs will still play back audio on a monaural TV but you would be wasting your money without the proper speakers and amplifier. This special audio feature is available on Beta hi-fi machines. On VHS machines it's called high definition (HD) audio.

Linear track audio is a standard audio feature. This record and playback method uses stationary audio heads to record sound with traditional recording techniques. The recording process is identical to methods used on reel-to-reel and audio cassette recorders. Some VCRs offer stereo linear audio. To enjoy stereo reproduction on linear audio heads you must connect an external amplifier and speakers.

Current VCR options include stereo tuners that let you record TV programs broadcast in stereo. This option should be considered if you'd like to hear stereo audio. VCRs that don't contain the stereo tuner, yet have the stereo audio option will play back prerecorded stereo programs and record stereo audio through the external audio input jacks.

Number of video heads is yet another option. VCRs have two to seven video heads mounted on the video head disk. Only two video heads are used at any given time to record and play back picture information. These two heads are mounted exactly one hundred and eighty degrees apart. One pair of heads may be used to record and play back at standard speed while a second set of heads can be used to record long play and extended play (super long play). Most four-head VHS machines work this way. Inexpensive VCRs use two heads to record and play back all three speeds.

If your VCR has high definition audio, then a separate pair of heads is used to record special high definition audio information. It is possible to have four video heads and two high definition audio heads mounted on the video head assembly. These machines have been called *six-head* VCRs.

Some machines have a special video head that is used only for improving special effects during play back. This head is used with the normal video heads during play back and is electronically switched on and off to provide optimum video presentation during slow motion and still frame modes. You can find VCRs with three, five, or seven heads mounted on the rotating head disk assembly.

Beta technolgy does not require as many heads. Beta hi-fi and video are recorded using the same video heads. It's typical to find Beta video recorders with Beta hi-fi and superb playback special effects using only three or four heads.

In VHS machines, the number of video heads can have a direct effect on the quality of playback special effects. VHS recorders with four video heads generally provide better picture quality in Standard Play than machines having only two heads recording in the same speed. This is because the extended play heads must be physically smaller to allow the recording information tracks to be spaced closer together on the tape. While the heads are smaller, the video head gaps are the same size.

The FM information track recorded onto the tape is not as wide as the tracks recorded by the larger SP heads so the FM signal being read off the tape is not as strong as the signal recorded with the SP heads. This reduces overall video playback quality. Four-head VHS machines with one set of heads for SP provide better, cleaner, video quality in this speed. Smaller heads are used for both the EP (SLP) and LP speeds.

VHS units with two video heads must use the smaller extended play head size for all three heads. The SP playback video is not as crisp on these units as it is on VCRs with four heads. One advantage of the four-head VHS VCR is the option to record and play back in standard speed for better video quality.

Some disadvantages of machines that have more than two video heads include increased replacement cost and potentially increased video tape wear because more heads contact the tape during each head rotation. Some VHS manufacturers recommend the replacement of both lower and upper drum assemblies during video head replacement on high definition audio capable VCRs because high precision is required for placement of the upper drum assembly. This creates a substantially higher cost than for the non-high definition machine. When video heads wear out on a standard VCR, only the upper video head disk needs to be replaced. Video heads cannot be replaced individually. The entire head drum must be replaced at one time.

A long list of options are available to the video recording buff. A typical advertisement today describes a VHS 14-day, four-event, 11-function infrared wireless remote, 105-channel cable-compatible VCR with Dolby Stereo sound. Sony recently marqueed their "SuperBeta Theatre" tabletop VCR with SuperBeta HiFi 85 dB FM audio/video recording capability. The selection process is getting more complicated.

If your machine is a second purchase unit, you may want to buy a different format VCR so you can play back both Beta and VHS. Thirty-seven percent of teenagers in the United States have their own VCR. Is the VCR to be used by one of your children? RCA claims that two of every 12 VCRs sold are now second purchases. Forty percent of households in the United States will own a VCR in 1987. Will you be trading tapes with your neighbor across the street? Will his machine be compatible with yours?

This is a sampling of the questions you should ask before purchasing a VCR. Other good sources of questions and "food for thought" are the consumer publications. Consumer Reports buying guides usually have a section rating VCR brands. By understanding yourself, your desires, the desires of those around you, and available products, you should be able to select a good VCR.

And VCRs are now being made in North America. Matsushita Electrical Industrial Company of Japan has a lease to manufacture VCRs and TVs in Canada. This will make parts easier to obtain and VCRs (hopefully) less expensive to repair.

CHOOSING THE BEST VIDEO TAPE

Technicians and engineers are often asked which brand of videotape is best to use in a VCR. From a technical perspective, one would guess that the best tape would have the thickest oxide layer, the strongest oxide binder, and the best head lubricant. While these factors are indeed important, the cassette housing itself is most critical in tape selection.

Magnetic tape production has reached such a level of maturity that many companies can successfully produce high quality tape. However, not all can produce high quality cassettes. That black, hard plastic tape housing that you take so much for granted, is produced with unbelievable precision. Should the reel tables not fit exactly into the cassette, the mechanics of the VCR could actually be damaged by the cassette.

Consumer Reports and Video Review magazines conducted lengthy studies on video cassette tapes and the results are interesting. Of prime importance is the finding that the range of quality between the best and worst tape is not large. Tape quality has improved greatly since 1981. Today, most video tapes can provide a better picture than a typical VCR can record or play back.

Manufacturers have introduced as many as a

half dozen grades of tape with as many price tags, but the difference in performance is slight. Another finding is that much like the "plain Jane" or "vanilla" floppy disks of the computer industry, the video cassette industry also has its no name, unlicensed tape vendors. If the tape does not meet the minimum quality standards of Beta or VHS, avoid these discount tapes. Your video tape is actually part of the equipment in your system so shop carefully. You could spend too much money and get too little quality for the price.

VIDEO PERFORMANCE

About two dozen companies manufacture video cassette tapes. Tape quality can be determined by analyzing its video and audio performance. Video performance can be broken into four categories: (1) dropout; (2) frequency response; (3) noise; and (4) signal retention.

Dropout

This is the most visible (and irritable) problem in videotape. Dropouts appear as intermittent horizontal streaks, tiny spots, glitches, or unwanted lines on the screen image. They are caused by missing bits of magnetic oxide, scratches on the tape surface, or dust particles that cling to the tape surface. This problem is measured by the number of dropouts that occur per minute of tape play. VHS tapes cannot exceed 50 fifteen microsecond periods of dropout in a minute of play. Most tapes average around 10 dropouts of 15 microseconds per minute.

Usually, dropout isn't noticed by the consumer-viewer because most VCRs have built-in circuitry that fills in the image holes when dropout occurs. In addition, vivid colors and large screen objects distract the viewer from noticing minor dropouts. Dropouts are most noticeable during slow-motion and freeze-frame.

Any video signal level reduction of 18 dB or more is considered a dropout. Dropouts tend to be worse at both ends of a videotape. This is why it's a good idea not to record anything on the first and last minute of a cassette tape.

Frequency Response

Frequency response is the same as picture resolution. Good frequency response produces good picture resolution. The better the frequency response, the better the definition, or detail in the playback picture (especially at high frequencies).

The tape output should have the same strength at all frequencies, but the VCR in playback amplifies the middle frequencies more. This isn't usually a problem during playback, except some tapes don't maintain the same signal output level at the highest frequencies and experience some loss in detail.

Video Noise

Video noise exists in two forms—luminance (brightness) noise and chrominance (color) noise. Luminance noise is seen as graininess or snow in the screen image. Chrominance noise can be noticed as amplitude modulation snow along horizontal lines of a color picture or as phase modulation changes in the shade of colors. At its worst, color noise can turn solid red areas into a confused disarray of blips and streaks.

Both luminance and chrominance noise are expressed as signal-to-noise ratios (SNR) and measured in decibels (dB). An SNR difference of 1 dB is perceptible by the viewer. The higher the SNR value, the better the quality of the tape. A luminance SNR of 45 dB or more is quite good. A chrominance SNR of 45 dB or better is very good. A value of 43 dB is noticeable. Most tapes demonstrate little luminance and chrominance noise (SNRs are high). In fact, the chrominance SNR for most videotapes is better than that of the consumer VCR.

Signal Retention

This is a measure of the longevity of a recorded program. Magnetic particles tend to forget their pattern after a number of plays. This lessening of the composite (RF) signal strength causes a signal loss that degrades the picture image. Since most of the degradation in image quality from demagnetization occurs after the first few times a tape is played, signal loss is measured in dB and is based on the loss in signal strength from first playback to after the tenth playback. A value of 3 dB represents a reduction in signal level by one half.

Most video tapes have less than 1 dB loss after ten replays.

AUDIO PERFORMANCE

Audio performance is determined by analyzing four factors: (1) tape edge regularity; (2) harmonic distortion; (3) dynamic range; and (4) bandwidth.

Tape Edge Regularity

This factor becomes especially important when you are using tapes in a stereo VCR. This type of VCR records two tracks in the same tape space area as is used for a single track of monaural audio. Conventional stereo VCRs record both right and left channel audio into this narrow band. The right channel touches the tape edge so damage near this edge results in a difference in signal strength between the right and left channels.

Video cassette tapes usually have no irregularities along the edges. However, improper storage of the cassette or misalignment in the tape path can damage the tape edge and dramatically affect the stereo playback.

New hi-fi VCRs record both sound and video diagonally across the tape so tape edge condition doesn't affect the sound quality.

Harmonic Distortion

The introduction of extraneous tones into the recording of a pure tone (usually 1 kHz) is called *harmonic distortion*. This measures the distortion the tape itself adds to an audio track. Harmonic distortion is expressed in percent and represents the amount of distortion of an audio signal recorded at 10 dB lower than a maximum (0 dB) reference. A distortion of one percent from a reference tone level can be detected by most listeners. Usually, no significant differences in harmonic distortion exist between tapes.

Dynamic Range

Dynamic range represents the span between the loudest and softest audio signal that can be played from a recorded tape. At the loud end, the signal gets badly distorted; at the soft end, the audio becomes noise or hiss. Dynamic range is measured in SNR and expressed in dB. Like video SNR, the higher the reading, the better. A value of 50 dB or more is required to qualify as high fidelity. Most videotapes cannot match the dynamic range of a good audio tape.

Bandwidth

This is an expression of the range of audio signals that a tape can reproduce. Most magnetic tape can record bass (midrange) tones so one way to determine bandwidth is to compare the recording of a high treble tone (Consumer Reports uses 11 kHz) to a midrange tone (1 kHz). The best tapes will reproduce the treble tone almost as well as the midrange tone. Video Review records 20 Hz to 20 kHz at 10 dB below a reference 0 dB level. During playback, they note the bass and treble frequencies where signal strength falls off by 3 dB. This is a half-power point where sound intensity loss can be clearly discerned.

With all these factors fresh in your mind, the only other factor yet to be decided is the format of your VCR. Battle lines have again been drawn in the war of video formats. The initial skirmish between Beta and VHS has ended with VHS the victor, but the battle is far from over. SuperBeta quality is superior to VHS and sales of machines with this capability are growing. Yet from another corner comes 8mm video with its younger and more advanced 8mm technology. The advantages of 8mm have been seen in home movie applications and the camcorder (camera-recorder). Samsung (Korea) recently demonstrated the Translator VCR with one recording deck for VHS and another for 8mm tape. With this machine, consumers can make 8mm home movies with a camcorder and then make VHS copies of those movies to send to relatives.

PREVENTIVE MAINTENANCE RECORD

Use this chart as a guide in conducting preventive maintenance on your VCR.

Maintenance Record

Date: _____

INITIAL CHECK

Playback _____ Fast Forward _____
Record _____ Rewind _____
Channel Tune _____ Threading _____

INTERMEDIATE CHECK AND REPAIR

CLEAN: **CLEAN OR REPLACE:**

Audio Heads _____ Belts _____
Video Heads _____ Tires _____
Control Track _____
Scanner _____
Tape Path _____ **LUBRICATE:**
Pulleys _____
Pinch Roller _____
Capstan _____ Reel Tables _____
Reel Tables _____ Idler Tires _____
Brake Pads _____ Pulley Shafts _____
Tension Bands _____

FINAL CHECK

Brake Torque _____ Record _____
FFWD. Torque _____ Playback _____
RWD. Torque _____ Tracking _____
Play Torque _____ RF Envelope _____
Hold Back Tension _____

Notes:

ROUTINE PERIODIC MAINTENANCE SCHEDULE

The following chart describes the suggested periodic maintenance and replacement actions which should be conducted on a VCR to achieve optimum performance and maximize operational life. In this chart, "C" stands for "clean," "L" stands for "lubricate," "O" stands for "oil," and "R" stands for "replace."

Suggested Periodic Maintenance and Replacement Schedule

DESCRIPTION	HOURS OF OPERATING TIME					
	500	1000	1500	2000	3000	5000
LOADING MOTOR	(Replace after 50,000 operations)					
CAPSTAN MOTOR				R		
VIDEO HEAD DRUM MOTOR						R
REEL TABLE MOTOR					R	
CASSETTE TRAY MOTOR (Front load units only)	(Replace after 20,000 operations)					
BELTS	C	R				
RUBBER TIRES	C	O,R				
RUBBER TIRE SHAFTS	C,O					
VIDEO HEADS	C		R			
STATIONARY AUDIO HEAD (Stack)	C			R		
FULL TRACK ERASE HEAD	C					
REEL TABLES	C	C,O				
BRAKES	C	R				
FWD. HOLDBACK TENSION BAND		R				
SENSOR LAMP (Incandescent only)		R				
TORQUE LIMITER ASSEMBLIES	C,O		R			
BELT PULLEYS AND SHAFTS		C,O				
MECHANICAL LINKAGES				C,L		
LOADING POSTS SLIDER BASE				C,L		
PINCH ROLLER	C	O		R		
VIDEO TAPE PATH GUIDES	C					
AUDIO HEADS	C			R		
CAPSTAN SHAFT	C	C,L				

Appendix **239**

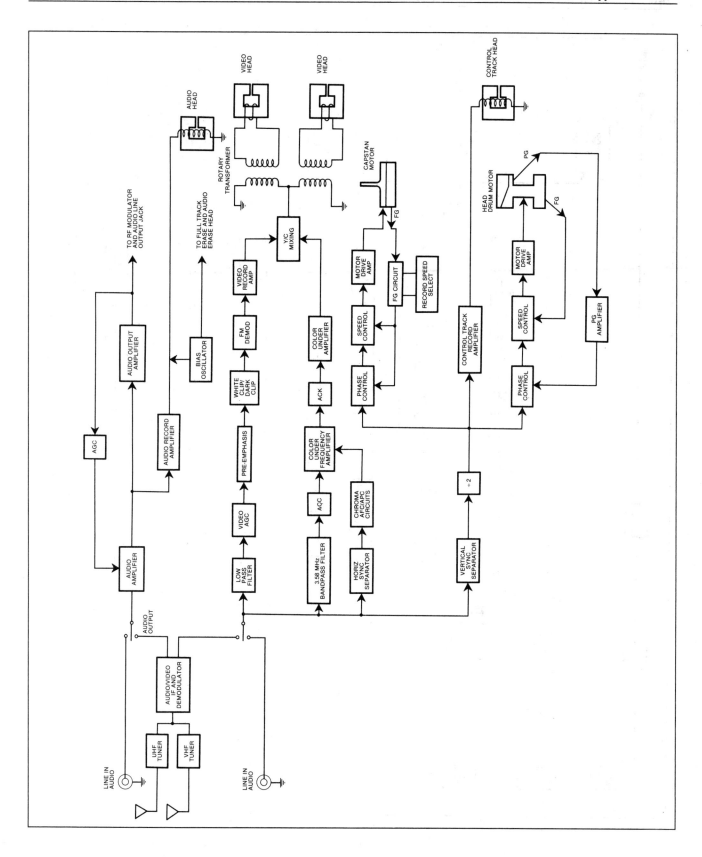

240 VCR TROUBLESHOOTING & REPAIR GUIDE

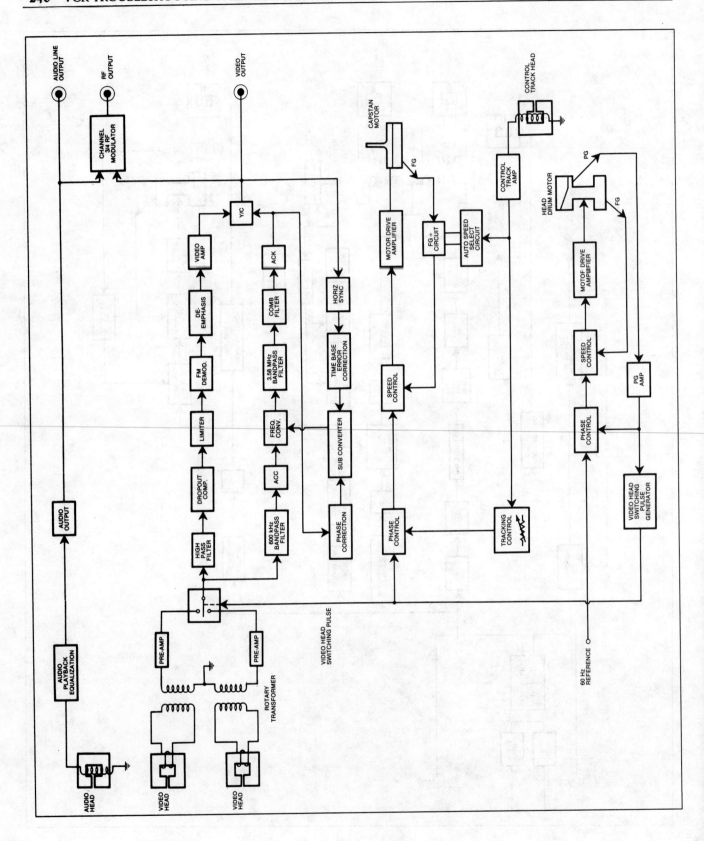

Index

A

A/B switch, 191
ACC. *See* Automatic color control
ACK. *See* Automatic color killer
AC volt meter (ACVM), 198
Adjustments, making, 39
AFC. *See* Automatic frequency control
AFM. *See* Audio frequency modulation
AGC. *See* Automatic gain control
Air circulation, importance of, 14
Alignment(s)
 gauges and jigs, 203-4
 making, 39
 tapes, 150, 205
Ampex Corporation, 2
Amplifier
 FM record, 148-49
 pre-, 103
 video playback pre-, 150-51
Amplitude, errors in, 145
AND gate, 131, 132
APC. *See* Automatic phase control
Arco Industries, 2
Atmospheric corrosion, effects of, 35
Audio/control head, 114
Audio erase head, 113-14, 115-16
Audio Frequency Modulation (AFM)
 creation of, 183
 crosstalk and, 183-84
 definition of, 50, 180
 head switching and, 185-86
 playback and, 185
 recording, 184-86
Audio performance, determining, 236
Audio problems
 audio from previous program still heard, 57-58
 audio plays back at wrong speed, 62

Audio problems—cont.
 buzz or hum in audio playback, 58
 low audio level in playback of self- and pre- recorded tapes, 61
 low audio level in playback of self-recorded tapes, 59
 no audio during playback, 55
 no audio or video in playback, 62
 playback volume level fluctuates, 54
 static or popping sound in hi-fi audio, 63
 won't record, 52
 wow and flutter, 57, 182-83
Audio recording techniques
 audio frequency modulation, 180, 183-88
 linear, 180-81, 233
 linear playback, 181-82
 specifications, 182-83, 188
Audio-video demodulators, tuners and, 193
Automatic color control (ACC), 135-37
 detect circuit, 140
 in VHS machines, 144
Automatic color killer (ACK), 137-38, 140-41
Automatic frequency control (AFC), 122, 123, 141-42
Automatic gain control (AGC), Y signal, 144-46
Automatic phase control (APC) circuitry, 123, 131, 139
 APC-ID, 139, 142
 detect circuit, 140, 142
Automatic speed selection, 170-71
Azimuth recording, 105-6, 124

B

Balanced modulator, 184

Bandwidth, 201, 236
Belts, changing, 43, 44, 45, 233
Beta Hi-Fi, 180
Betamax systems, 3
 automatic color control in, 135-37
 automatic color killer in, 137
 automatic frequency control in, 141-42
 Beta I, 162
 Beta II, 162
 chroma crosstalk, 126, 128-30
 color playback processing circuits in, 139
 color subcarrier and, 120-21
 drum servo in, 158-62
 free running frequency in, 147
 frequency modulation in, 144
 head sizes of, 105, 111
 playback servo control in early, 161
 record capstan servo in early, 160-61
 recording in, 180
 servo error correction in early, 160
 sidebands in, 102
 subcarrier phase relationship in, 127-28
 SuperBeta, 156-57
 tape leaders in, 176
 tape paths for, 112
 tape speed for, 99
 writing speed in, 101
BH curve, 93-94
Bias, 95
BID. *See* Burst identification circuit
Bing Crosby Enterprises, 2
Blackouts, 33-34
Blown fuses, 17
Blown-up devices, 13
Brownouts, 33
Burst, 101, 119
 boost, 142, 144

241

Burst flag, horizontal synchronization pulse, 123, 136
Burst gate circuit, 123, 136, 138, 139
Burst identification circuit (BID), 138-40

C

Cable connectors, problem with, 14
Cable-ready tuners, 189-90, 232
Capacitance measuring, 211
Capacitors
    electrolytic, 10, 207
    failure of, 12, 207
    film, 207
    Mylar, 10, 11
    replacing, 211
    Tantalum, 10
    testing, 210-11
    variable, 10, 11
Capstan motor, defective, 14
Capstan servo, 110, 158
    early Beta record, 160-61
    phase lock, 169
    playback, 169
    purpose of, 167-69
    special effects and, 172-73
Carbon film device resistor, 9
Carrier leak, 148
Carrier ripple, 148
Cartridge machines, development of, 2-3
Cartrivision, 2-3
Cassette-down sensors/switches, 16
Cassettes
    cleaning, 40-42
    how to select, 235-36
    test, 16
Cassette tray assembly, 15
Ceramic Film capacitors, 10, 11
Chips. See Integrated circuits
Chroma/color subcarrier signal, 119
Chroma crosstalk, 12
    audio frequency modulation and, 183-84
    Betamax, 126, 128-30
    VHS, 132-33
Chroma phase selection circuit, 126, 129
Chroma signal processing, record and, 140-42, 144-47
Chroma synchronization pulses, 118
Chroma time base errors, 119, 142
Chrominance noise, 235
Circuit boards, 8
    repairing, 216-17
Cleaning
    cassettes, 40-42
    frequency of, 43
    heads, 39-40, 42-43
    manually, 42-43
Clipping, white and dark, 147
Clocks, microprocessors and, 178
Clock/timer circuitry, 175
Clock/timer problems
    clock loses 10 minutes each hour, 89
    no or intermittent record using timer, 90
Coersive force, 94
Cold solder joints, 216
Cold temperatures, effects of, 24-25
Color/chroma subcarrier signal, 119
Color frequencies, recording of, 119
Color problems
    bands of color on screen during color playback, 75
    flicker, 136, 138

Color problems—cont.
    no color in record or playback, 73
    no color on newly recorded tapes, 71
Color under, 119-21, 124
    automatic phase control and, 139
    subcarrier phase, 128
    up-conversion of, 141
Comb filters, 125-26, 129-30
    as low pass filters, 144
Consumer video (CV), introduction of, 2
Control track pulse/signals, 95-96, 160, 169-70
Cooling techniques, use of, 210
Corrosion, 35-37
Counter electromotive force (EMF), 158-59
Cross modulation, 105
Cross pulse monitor, 103, 199-200
Crosstalk
    audio frequency modulation and, 183
    azimuth recording and, 105-6, 124
    in Betamax, 124, 126, 128-30
    chroma, 12, 183-84
    description of, 104-5
    in VHS, 132-33
Crystal ringer circuit, 136
Cue-and-review function, 110-11, 174
Cutters, 217
CV. See Consumer video
CW source, 126-34

D

DC information, 102
Decibels (dB), 97
De-emphasis, 156
Delay line input transducer, 124-25, 129
Demodulators
    frequency modulation and, 103, 154, 155
    tuners and audio-video, 193
Depth multiplexing, 187-88
Dew sensor, 25
Dielectric, 207
Digital-mulitmeter (DMM), 198
Digital phase-locked-loop (D-PLL), 142
Digital servo processing, 171-72
Digital-volt-meter (DVM), 198
Dihedral error, 118
Diode effect, 212
Diodes, 11
    failure of, 12, 208
    testing, 211-12
DIP. See Dual-in-line package
Direct drive video head drum motor, 163
    advantage of, 232
    servo circuits for, 165-66
    speed and phase control for, 163-65
Direct oxidation, effects of, 35
DMM (digital-mulitmeter), 198
Doubling circuit, 154
Dropout, 139, 235
    compensation (DOC), 151-56
Drum record phase, 160
Drum servo, Beta, 158-62
Dual-in-line package (DIP)
    extractor tool, 214
    network resistor, 10
Dust and foreign particles, controlling, 25-27
DVM (digital-volt-meter), 198
Dykes, 217
Dynamic range, 98, 101, 236

E

Eccentricity gauge, 204
Eddy currents, 93
E-E (Electronic-to-Electronic), 50, 156
Electrolytic capacitors, 10, 207
Electro-magnet, 158-59
Electromagnetic interference (EMI), 27, 29-30
Electromotive force (EMF), counter, 158-59
Electronic Industries Association of Japan (EIAJ), 2
Electrostatic discharge (ESD), 27, 30-31
Equalization, 97-98
Extended play (EP) speed, 107-8

F

Ferrite, 93
FG. See Frequency generator
Film capacitor, 207
Filters, power problems and, 35
Flagging, 69, 118, 119, 203
Flip-flop, 138-39, 161
Floating, 209
Flutter, 57, 182-83, 188
Flux, 91, 92, 215
    density, 94
FM. See Frequency modulation
Free running speed, 158
Frequency correction, 123-24
Frequency counter, 201
Frequency generator pulses, 165
Frequency generators (FGs), 200
Frequency interlace, 126
Frequency modulation (FM), 98
    demodulator, 103, 154, 155
    luminance, 144
    modulator, 147-51
    record amplifier, 148-49
    recording/playback, 101-4
Frequency response, 182, 188, 235
Fried, capacitors and, 207
Full field recording, 2

G

Galvanic corrosion, effects of, 36
Generators
    frequency, 200
    NTSC, 198-99
    pulse, 131, 165
    ramp, 160
    sweep, 149, 200
Grease, use of, 44
Guard bands, 104-6

H

HAFC. See Horizontal automatic frequency control
Hall ICs, 163-65
Handshaking, 177
Harmonic distortion, 182, 236
HD (High Definition), 180, 186-87, 233
Head drum motor, direct drive video, 163
    speed and phase control for, 163-65
Head drum playback servo, 171
Head drum servo, purpose of, 166
Head gap, 98-104
Heads
    audio/control, 114

Heads—cont.
  audio erase, 113-14, 115-16
  azimuth recording and, 105-6
  cleaning of, 39-40, 42-43
  construction of, 91-93
  number of, 233-34
  size of, 99, 107-8, 110
  stationary, 98
  switching of, 103-4
  time base error and misaligned, 118
  use of four, 107-8
  use of two, 99-100
Head switching, audio frequency
    modulation and, 185-86
Heat, effects of, 24
Heating techniques, use of, 210
Helical scan, 2
Helical-scan recording, 99
Heterodyne mixer circuit, 136, 137-38, 140
Heterodyne submixer circuit, 126
High Definition (HD), 180, 186-87, 233
High density recording, 106-8, 110-16
High quality (HQ), VHS, 157-58
Horizontal automatic frequency control
    (HAFC), 119
Horizontal synchronization pulses, 101, 118,
    126, 138
Humidity, effects of high, 118
Hysteresis AC motor, 158
Hysteresis loop, 94

I

ICs. See Integrated circuits
Improper operation or no functions
  improper rewinding of tape, 81
  mode button won't engage on first try,
      88
  no PLAY, FAST FORWARD, REWIND, or
      RECORD, 84-86
  no rewind at all, 83
  only eject works, 79
  rewind stops before end of tape, 82
  shuts down or returns to STOP after a
      few seconds, 80
  slow or no FAST FORWARD, 88
  tape spills into VCR, 87
Inductors, 11
Input/output expander (I/O), 177-78
Input section, 4
Insert edit modes, 149
Integrated circuits (ICs), 8-9
  bent or broken, 13
  failure of, 206-7, 210
  flat pac, 214
  Hall, 163-65
  removing and replacing, 18-19, 171
Intermittent problems, 18
Isolated variable AC supply, 201
Isolators, power problems and, 34-35

J

JVC, 2, 3

L

Leaking, failure due to, 12
Levers, bent, 14, 15
Limiting circuit, 151, 152
Linear audio recording, 180-81

Linear playback audio, 181-82
Linear track audio, 50, 233
Liquids, spilled, 13-14
Litton Systems, Inc., 30
Long play (LP), 142, 144
Luminance, 101, 119, 235
  frequency modulation, 144

M

Machine oil, use of, 43-44
Magnetic fields, 37, 91
  intensity, 93-94
Magnetic saturation point, 94
Magnetic tape
  description of, 4-5
  recording principles of, 93-98
Magnets, use of, 160, 163
Maintenance, preventive
  corrosion, 35-37
  dust and foreign particles, 25-27
  magnetic fields, 37
  moisture, 25
  noise interference, 27-33
  power line problems, 33-35
  record of, 237-38
  tape drive mechanism and, 43-46
  temperature extremes and, 24-25
  VCRs and, 39-47
  video tapes and, 37-39
Matsushita Electrical Industrial Company, 2,
    234
Mechanical approach to troubleshooting, 15
Mechanical chassis assembly, 15
Memory stop, 176
Metal migration, 206
Metal oxide varistor (MOV), 34-35
Meters, 197-98
Microprocessors
  failure of, 12
  pins in, 179
  purpose of, 174-76
  workings of, 177-80
MMV. See Monostable multivibrator
Modulators
  balanced, 184
  frequency, 147-51
  radio frequency, 193-94
Moisture, controlling, 25
Monaural audio, 180
Monitor, cross pulse, 103, 199-200
Monitors/receivers, 199
Mono audio, 180
Monostable multivibrator (MMV), 154-55,
    171
MOV. See Metal oxide varistor
Mylar capacitors, 10, 11

N

Noise band, 110-11
Noise interference, 17-18, 27-29
  chrominance, 235
  controlling, 29
  effects of, 17-18
  electromagnetic interference, 27, 29-30
  electrostatic discharge, 27, 30-31
  hiss, 98
  luminance, 235
  radio frequency interference, 27, 31-33
  sources of, 27-29

NTSC (National Television Systems
    Committee) generator, 198-99
Nut drivers, metric, 217

O

OR gate, 132, 138-39
Oscilloscope, 201-2, 217

P

Panasonic
  head sizes of, 111
  PV 1100, 106
  PV 5000, 108, 111
Pause, 172
  limit function, 176-77
Peak detection, 144
Pedestal level, 146-47
PG. See Pulse generator
Phase comparison circuit, 123, 131
Phase control unit, 158
  for direct drive motor, 163-65
Phase error corrections, 123-24
Phase inverter selection circuit, 126, 127,
    138-39
Phase rotation circuitry, 131
Picture brightness, 145
Playback
  audio frequency modulation and, 185
  chroma level, 136
  distortion, 97
  head drum playback servo, 171
  linear playback audio, 181-82
  phase restoration, 127-28
  pre-amplifier, 150-51
  servo (capstan), 169
  servo control in Beta, 161
  slow motion, 172, 173
  subcarrier phase and, 138
Pliers, 217
Polarities, 92, 93
Poniatoff, Alexander, 2
Power line problems, 33-35
Power supply, 194-95
Power supply problems
  no functions, 76
  no modes functional, except eject
      works, 78
  no power (dead VCR), 77
Pre-amplifier, 103
  video playback, 150-51
Pre-emphasis, 146-47
Pulse generator (PG)
  enabling circuit, 131
  pulses, 165
Pulse width modulator (PWM), 172

R

Rabbet, 113
Radio frequency (RF)
  modulators, 193-94
  sweep generator, 149, 200
Radio frequency interference (RFI), 27,
    31-33
Radio frequency output waveform
    envelope, 108
Ramp generator, 160
RCA Corporation, 2
Reader section, 4

Ready/acknowledge line, 177
Record amplifier, FM, 148-49
Record capstan servo, early Beta, 160-61
Recorded wavelength, 95
Recorder section, 4
Recording
    chroma signal processing and, 140-42, 144-47
    of color frequencies, 119
    limitations of video, 93-95
    linear audio, 180-81
Record phase, drum, 160
Reel table motor, 174
Regulators, power problems and, 35
Reluctance, 93
Remote control transmitters, 175
Repair center(s)
    costs charged by, 5, 23
    how to prepare for and select a, 19-21
    when to call a, 50
Repair parts, spare, 218
Repairs, common, 50
    how to make, 208-10
    use of heating and cooling techniques, 210
    See also Tools for making repairs
Resistors, 9-10
    damping, 150
    failure of, 12, 207-8
Resolution, increasing, 125
Retention, 92
RF. See Radio frequency
RFI. See Radio frequency interference
RF switch, 191
Rotary transformer, 149-50
Rotational phase converter, 132-33

S

Sample-and-hold circuit, 160, 161, 169
SAMS Photofacts, 50-51
Scanner, 99
Scan pulses, 179
Scope, 201-2
Screwdrivers, metric, 217
Search function, 110-11
Sears & Roebuck, 3
Servo control circuit, 14, 158
    Beta drum servo, 158-62
    Beta playback, 161
    Beta record capstan, 160-61
    capstan, 167-69, 172-73
    capstan playback, 169
    digital technology and, 171-72
    for direct drive motors, 165-66
    head drum, 166
    head drum playback, 171
    for multiple speed machines, 162-66
Servo error correction, early Beta, 160
Set-up, video, 146-47
Sidebands, 102
Signal retention, 235-36
Signal-to-noise ratio (SNR), 235
Single channel audio, 180
Single-in-line package (SIP) network resistor, 10
Skip field recording, 2
Slow motion playback, 172, 173
SLP. See Super long play
Snow bands, 67, 111, 169
SNR (signal-to-noise ratio), 235
Solder
    cold solder joints, 216

Solder—cont.
    extra, 13
    how to, 215-16
    how to remove, 213-14
    sucker, 213
Soldered parts, repairing, 17
Soldering iron, 215-16, 217
Sony, 2
    audio frequency modulation and, 183
    audio recording method of, 180
    Betamax, 3
    Beta I, 162
    Beta II, 162
    SuperBeta, 156-57
    U-Matic VCR, 2
SP. See Standard play
Speed and phase control for direct drive motor, 163-65
Splitter, 191
Standard play (SP), 142, 144
Stereo audio, 180
Still frame, 172, 173
Stop
    automatic, 176
    memory, 176
    sensors, 176
Storage medium, 4
Subcarrier phase relationship
    in Betamax systems, 127-28
    playback and, 138
    in VHS systems, 130-33
SuperBeta, 156-57
Super long play (SLP) speed, 107-8, 142, 144
Surge protectors, 34
Sweep generators, 149, 200
Symmetry, 148
Synchronization pulses, 116
    burst flag horizontal, 123
    chroma, 118
    horizontal, 101, 118, 126, 138
    vertical, 101, 158
Sync tip clamp, 146
System control circuitry
    clock/timer functions and, 175
    failure of, 177
    functions of, 173-77
    operator input and, 174
    power supply and, 173-74
    reel table motors and, 174
    remote control transmitters and, 175
    special effects and, 174
    stop functions and, 176-77
    tape load and unload functions and, 174-75
    tape sensing features and, 176

T

Tank circuit, 150
Tantalum capacitors, 10
Tape(s)
    alignment, 150, 205
    construction of, 92
    dynamic range of, 98
    end of, 176
    how to select, 234-35
    leaders of, 176
    path, 112-16
    remaining indicators, 176
    tension, 116, 174
Tape drive mechanism, preventive maintenance for, 43-46

Tape edge regularity, 236
Technical (electronic) approach to troubleshooting, 15
Temperature extremes, 24-25
Tension and torques, adjusting, 46-47
Tension meters, 202-3
Tension of tape, 116, 174
Test & Measurement Systems Company, 201
Test equipment, types of, 15
Testing of
    capacitors, 210-11
    diodes, 211-12
    transistors, 212-13
Thermal stress, 12-13, 206
Time base errors
    chroma, 119, 142
    dihedral error, 118
    horizontal, 118
    reasons for, 116, 118
    TV horizontal automatic frequency control circuit and, 119
Timers, 232
    See also Clock/timer
Tinning, 216
Tires, 174
    replacing, 45-46
Tools for making repairs
    alignment gauges and jigs, 203-4
    alignment tapes, 150, 205
    cost of, 217
    cross pulse monitor, 103, 199-200
    cutters, 217
    eccentricity gauge, 204
    frequency counter, 201
    frequency generators, 200
    isolated variable AC supply, 201
    meters, 197-98
    metric nut drivers, 217
    metric screwdrivers, 217
    monitors/receivers, 199
    NTSC generator, 198-99
    oscilloscope, 201-2, 217
    pliers, 217
    soldering iron, 215-16, 217
    tension meters, 202-3
    torque gauges, 202
    Variac, 201
Torque, adjusting, 46-47
Torque gauges, 202
Toshiba, 2
Tracking, 161
    control, 170
Transients, 34
Transistors
    failure of, 12, 208
    testing, 212-13
Triboelectric Series, 30
Troubleshooting
    localizing failures, 15-19
    parts that usually fail, 13
    reasons for failure, 14-15
    safety precautions when, 21-22
    steps for, 7-8, 17
TV horizontal automatic frequency control circuit, 119
Tuners
    audio-video demodulators and, 193
    cable-ready, 189-90, 232
    hookup configurations, 190-93
    VHF, UHF, mid- and super-band channels, 188-89
Tweakers, 217
2 to 1 interlace, 100

Index

U

UHF, 188–89
U-Matic VCR, 2
Up-converter, 139
Upper cylinder/drum, 99

V

Vacuum-tube-voltmeter (VTVM), 197
Variable capacitors, 10, 11
Variable crystal oscillator (VXO), 140
Variable resistor, 10
Variac, 201
VCO. *See* Voltage controlled oscillator
Vertical synchronization pulses, 101, 158
VHF, 188–89
VHS (Video Home System), 3
   automatic color control in, 144
   chroma circuits in, 142
   color subcarrier and, 120–21
   crosstalk in, 132–33
   free running frequency in, 147
   frequency modulation in, 144
   high quality, 157–58
   recording in, 180, 186–87
   subcarrier phase relationship in, 130–33
   tape leaders in, 176
   tape path for, 112

VHS (Video Home System)—cont.
   tape speed for, 99
   writing speed in, 101
Video cassette recorder (VCR)
   components of a, 4
   development of, 2–3
   how to hook-up to a television, 4
   selecting a, 231–34
Video head drum motor, direct drive, 163
   speed and phase control for, 163–65
Video head drum servo, 158
Video head switching pulse, 127, 138, 151, 161–62
Video problems
   fine horizontal line floats through picture, 66
   no video or audio in playback, 68
   picture alternates between good video and snow, 67
   snow bands or lines of interference at top or bottom of screen, 70
   snow on screen during playback, 64
   vertical images bend or tear near top of screen, 69
Video stair step, 147
Video tape recorder (VTR) systems, development of, 1–2
Video tapes, preventive maintenance for, 37–39

Video tape wear, 14
Voltage controlled oscillator (VCO), 122–23
   frequency errors in, 140, 142
VOM (volt-ohm-millimeter), 197
VTVM (vacuum-tube-voltmeter), 197
VXO (variable crystal oscillator), 140

W

Wavelength
   how to calculate, 95
   importance of, 95–97
   recorded, 95
White and dark clipping, 147
Wow and flutter, 57, 182–83, 188
Writing speed, 98–101

Y

Y signal, automatic gain control and, 144–46

Z

Zero guard band, 105, 126

MORE FROM SAMS

☐ **How to Read Schematics (4th Edition)**
Donald E. Herrington
More than 100,000 copies in print! This update of a standard reference features expanded coverage of logic diagrams and a chapter on flowcharts. Beginning with a general discussion of electronic diagrams, the book systematically covers the various components that comprise a circuit. It explains logic symbols and their use in digital circuits, interprets sample schematics, analyzes the operation of a radio receiver, and explains the various kinds of logic gates. Review questions end each chapter.
ISBN: 0-672-22457-7, $14.95

☐ **Cable Television (2nd Edition)**
John Cunningham
With this text, engineers and technicians can learn to examine each component in a cable system, alone and in relation to the system as a whole. Sections include component testing, troubleshooting, noise reduction, and system failure.
ISBN: 0-672-21755-4, $15.95

☐ **Video Cameras: Theory & Servicing**
Gerald P. McGinty
This entry-level technical primer on video camera servicing gives a clear, well-illustrated presentation of practical theory. From the image tube through the electronics to final interface, all concepts are fully discussed.
ISBN: 0-672-22382-1, $18.95

☐ **Television Symptom Diagnosis (2nd Edition)** *by Richard W. Tinnel*
This easy-to-use text provides you with a basis for entry-level servicing of monochrome and color TV sets. It focuses on identification of abnormal circuit operations and symptom analysis.
ISBN: 0-672-21460-1, $15.95

☐ **John D. Lenk's Troubleshooting & Repair of Microprocessor-Based Equipment**
John D. Lenk
Here are general procedures, techniques, and tips for troubleshooting equipment containing microprocessors from one of the foremost authors on electronics and troubleshooting. In this general reference title, Lenk offers a basic approach to troubleshooting that is replete with concrete examples related to specific equipment, including VCRs and compact disc players. He highlights test equipment and pays special attention to common problems encountered when troubleshooting microprocessor-based equipment.
ISBN: 0-672-22476-3, $21.95

☐ **Know Your Oscilloscope (5th Edition)**
Robert G. Middleton, Revised by Joseph Carr
The oscilloscope remains the principal diagnostic and repair tool for electronic technicians. This book provides practical data on the oscilloscope and its use in TV and radio alignment, frequency and phase measurements, amplifier testing and signal tracing, and digital equipment servicing. Additional material is provided on oscilloscope circuits and accessories. A vital reference for your workbench.
ISBN: 0-672-22473-9, $16.95

☐ **The Sams Hookup Book: Do-It-Yourself Connections for Your VCR**
Howard W. Sams Engineering Staff
Here is all the information needed for simple to complex hook ups of home entertainment equipment. This step-by-step guide provides instructions to hook up a video cassette recorder to a TV, cable converter, satellite receiver, remote control, block converter, or video disk player.
ISBN: 0-672-22248-5, $4.95

Look for these Sams Books at your local bookstore.

To order direct, call 800-428-SAMS or fill out the form below.

Please send me the books whose titles and numbers I have listed below.

Name *(please print)* _____
Address _____
City _____
State/Zip _____
Signature _____
(required for credit card purchases)

All States add local sales tax _____

Enclosed is a check or money order for $ _____
Include $2.50 postage and handling.

Charge my: ☐ VISA ☐ MC ☐ AE
Account No. _____ Expiration Date _____

Mail to: Howard W. Sams & Co.
Dept. DM
4300 West 62nd Street
Indianapolis, IN 46268

DC115